AF614713

METHODS IN MOLECULAR BIOLOGY™

Series Editor
John M. Walker
School of Life Sciences
University of Hertfordshire
Hatfield, Hertfordshire, AL10 9AB, UK

For further volumes:
http://www.springer.com/series/7651

Embryonic Stem Cell Immunobiology

Methods and Protocols

Edited by

Nicholas Zavazava

Department of Internal Medicine and VAMC Iowa City, Division of Immunology
Roy J. and Lucille A. Carver College of Medicine, University of Iowa, Iowa City, IA, USA
Immunology Graduate Program, University of Iowa, Iowa City, IA, USA
Veterans Affairs Medical Center, Iowa City, IA, USA

Editor
Nicholas Zavazava
Department of Internal Medicine and VAMC Iowa City
Division of Immunology
Roy J. and Lucille A. Carver College of Medicine
University of Iowa
Iowa City, IA, USA

Immunology Graduate Program
University of Iowa, Iowa City, IA, USA

Veterans Affairs Medical Center
Iowa City, IA, USA

ISSN 1064-3745 ISSN 1940-6029 (electronic)
ISBN 978-1-62703-477-7 ISBN 978-1-62703-478-4 (eBook)
DOI 10.1007/978-1-62703-478-4
Springer New York Heidelberg Dordrecht London

Library of Congress Control Number: 2013940269

Printed on acid-free paper

Humana Press is a brand of Springer
Springer is part of Springer Science+Business Media (www.springer.com)

Preface

Bone marrow stem cells are the most transplanted cells worldwide. These cells are used as a replacement therapy for patients suffering from a diverse number of hematopoietic diseases and immunodeficiencies. In contrast, the use of bone marrow cells in regenerative medicine has so far remained without much success. For example, several studies have been published on the inability of bone marrow cells to repair acutely infarcted cardiac tissue [1–4]. In addition, bone marrow cells are highly immunogenic, requiring harsh immunosuppressive regimens to prevent rejection and graft-versus-host disease. This limits the number of bone marrow recipients to a small number of patients that are HLA compatible. In the new era of pluripotent stem cells, great opportunities for establishing new therapies have opened up. The discovery of human embryonic stem cells [5] and that of induced pluripotent stem (iPS) cells [6] has made it possible to derive any desired tissues for regenerative medicine. In contrast, iPS cell-derived cells are only limited by the lack of established protocols that can be applied in humans. The derivatives of pluripotent stem cells lack MHC class II antigens and poorly express class I antigens, which is an advantage when transplanting the cells across MHC barriers. Dr. Shinya Yamanaka, who first described iPS cells, was recently awarded the 2012 joint Nobel prize for Physiology and Medicine for this pioneering work [6–8].

Clearly the challenge is to establish new protocols that allow the successful differentiation of these cells into lineage committed cells. A lot of progress has now been made in mouse cells. For example, it is now feasible to efficiently derive hematopoietic cells from mouse ES cells [9–11]. These cells have been shown to expand in vitro and to engraft long-term. The caveat is that it has so far not been possible to show multi-lineage commitment and functional studies of ES cell derivatives. So far, few studies have been reported on the function of iPS cell derivatives. For example, ES cell-derived T cells were shown to respond to viral antigen. Tetramer staining and intracellular staining of T cells showed response to antigen stimulation by the T cells [11, 12]. Functional studies on iPS cell-derived cells are even more limited. Hanna et al. reconstituted peripheral blood of sickle cell anemic mice with hematopoietic progenitors derived from iPS cells where the sickle cell anemia gene had been corrected. This study for the first time showed the potential of iPS cells in regenerative medicine. Since the cells are derived from self, there is no requirement for immunosuppression. However, the cells were targeted by NK cells, although they were from "self." We recently reported that ES cell-derived hematopoietic cells hardly express MHC antigens, making them highly susceptible to NK cells [13].

In this book, a variety of topics are discussed. In particular, hematopoietic cells derived from ES cells are tackled by a number of authors. The interaction of these cells with natural killer cells or with cytotoxic T cells will be discussed as well. Additionally, a few chapters deal with the establishment of specific protocols for the derivation of hematopoietic cells and neuronal cells. This book offers the expert and nonexpert different aspects of stem cells. The content is very timely to the field as we inch closer to the use of stem cells derived

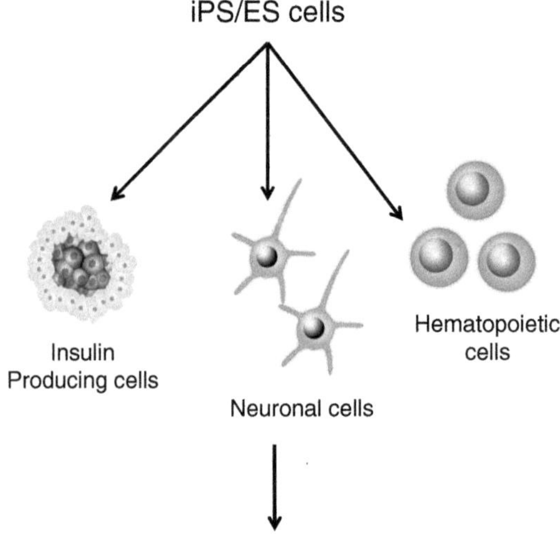

Fig. 1 Characterization of pluripotent stem cell derivatives: Pluripotent stem cells can be differentiated into insulin-producing cells, neuronal cells, and hematopoietic cells. The impact of T cells, NK cells, and the humoral response to these cells in the in vivo setting remains to be studied

from pluripotent stem cells to the clinic. The main challenges that remain are establishing robust protocols for the derivation of desired cells and establishing the immunological characteristics of these derivatives. For example, we recently established that ES cell-derived hematopoietic cells poorly express MHC antigens [11]. This characteristic makes the cells vulnerable to NK cells [13], but also makes them evade allogenic cytotoxic T cell deletion. This is supported by an earlier report by us that showed that transplantation of non-differentiated ES cells across MHC barriers leads to their partial differentiation into hematopoietic cells allowing the establishment of mixed chimerism and transplantation tolerance. These studies need to be extended with other ES cell derivatives extending our possible application of these cells into future therapies. The advantage of using patient-tailored iPS cells is that there is no anticipation of allogenic rejection of the cells and their derivatives. A possible immunological concern is that the derivatives of pluripotent stem cells could be susceptible to NK cell killing due to their low expression of MHC class I antigens. These concepts are summarized in Fig. 1.

An interesting area where pluripotent stem cells could make a huge difference in the design of new cell-based therapies is in type 1 diabetes. This disease is a result of the auto-destruction of pancreatic β cells. Therefore treatment could be established by replacing the destructed cells. In vitro, it has taken at least 10 years to establish more robust protocols for the derivation of pancreatic cells. An initial protocol established in the mouse turned out to be an artifact and irreproducible [14]. Fortunately, in the past 5–7 years studies on human ES cell-derived insulin-producing cells have made significant progress. Studies by the Baetge group [15–19] have made important contributions. The disadvantages of their approach are that the yield is low and the cells are not mature β cells to the field. A new approach that has now been published in both the mouse and the humans takes advantage of the generation of endodermal cells that can be purified and further differentiated into insulin-producing cells [20, 21]. In both protocols the transplanted insulin-producing cells normalized glucose in diabetic mice and survived long-term in vivo. In both protocols, full maturation of the insulin-producing cells takes 8 months in vivo. Despite the slow process in achieving robust insulin production, this approach could be a real alternative to the treatment of diabetes in humans.

There is no doubt that stem cells present a new and innovative platform for establishing new cell-based therapies. What we need to do is better define the immunogenicity of these cells and establish more efficient protocols for the derivation of the cell types of interest. We predict that rapid progress will be made in the future and that stem cell-based therapies will be established possibly within the next decade.

Iowa City, IA, USA *Nicholas Zavazava*

References

1. Lunde K et al. (2006) Intracoronary injection of mononuclear bone marrow cells in acute myocardial infarction. N Engl J Med 355:1199–1209
2. Menard C et al. (2005) Transplantation of cardiac-committed mouse embryonic stem cells to infarcted sheep myocardium: a preclinical study. Lancet 366:1005–1012
3. Orlic D et al. (2001) Bone marrow cells regenerate infarcted myocardium. Nature 410:701–705
4. Schachinger V et al. (2006) Intracoronary bone marrow-derived progenitor cells in acute myocardial infarction. N Engl J Med 355:1210–1221
5. Thomson JA (1998) Embryonic stem cell lines derived from human blastocysts. Science 282:1145–1147
6. Takahashi K, Yamanaka S (2006) Induction of pluripotent stem cells from mouse embryonic and adult fibroblast cultures by defined factors. Cell 126:663–676
7. Okita K, Ichisaka T, Yamanaka S (2007) Generation of germline-competent induced pluripotent stem cells. Nature 448:313–317
8. Aoi T et al. (2008) Generation of pluripotent stem cells from adult mouse liver and stomach cells. Science 321:699–702
9. Kyba M, Perlingeiro RC, Daley GQ (2002) HoxB4 confers definitive lymphoid-myeloid engraftment potential on embryonic stem cell and yolk-sac hematopoietic progenitors. Cell 109:29–37
10. Schiedlmeier B et al. (2003) High-level ectopic HOXB4 expression confers a profound in vivo competitive growth advantage on human cord blood CD34+ cells, but impairs lymphomyeloid differentiation. Blood 101:1759–1768
11. Chan KM, Bonde S, Klump H, Zavazava N (2008) Hematopoiesis and immunity of HOXB4-transduced embryonic stem cell-derived hematopoietic progenitor cells. Blood 111:2953–2961
12. Schmitt TM, Ciofani M, Petrie HT, Zuniga-Pflucker JC (2004) Maintenance of T cell specification and differentiation requires recurrent notch receptor–ligand interactions. J Exp Med 200:469–479
13. Tabayoyong WB, Salas JG, Bonde S, Zavazava N (2009) HOXB4-transduced embryonic stem cell-derived Lin-c-kit+ and Lin-Sca-1+ hematopoietic progenitors express H60 and are targeted by NK cells. J Immunol 183:5449–5457
14. Lumelsky N (2001) Differentiation of embryonic stem cells to insulin-secreting structures similar to pancreatic islets. Science 292:1389–1394
15. D'Amour KA et al. (2006) Production of pancreatic hormone-expressing endocrine cells from human embryonic stem cells. Nat Biotechnol 24:1392–1401
16. D'Amour KA et al. (2005) Efficient differentiation of human embryonic stem cells to definitive endoderm. Nat Biotechnol 23:1534–1541
17. Kelly OG et al. (2011) Cell-surface markers for the isolation of pancreatic cell types derived from human embryonic stem cells. Nat Biotechnol 29:750–756
18. Kroon E et al. (2008) Pancreatic endoderm derived from human embryonic stem cells generates glucose-responsive insulin-secreting cells in vivo. Nat Biotechnol 26:443–452
19. Schulz TC et al. (2012) A scalable system for production of functional pancreatic progenitors from human embryonic stem cells. Plos One 7:e37004

20. Raikwar S, Zavazava N (2012) PDX1-engineered embryonic stem cell-derived insulin producing cells regulate hyperglycemia in diabetic mice. Transplant Res 1:19

21. Rezania A et al. (2012) Maturation of human embryonic stem cell–derived pancreatic progenitors into functional islets capable of treating pre-existing diabetes in mice. Diabetes 61:2016–2029

Acknowledgments

This material is based upon work supported in part by the Department of Veterans Affairs, Veterans Health Administration, Office of Research and Development, Biomedical Laboratory Research and Development 1 I01 BX001125-02 and by NIH/NHLBI Grants 5 R01 HL073015-08 and 3 R01 HL073015-04A1S1. The content of this manuscript is solely the responsibility of the authors and does not necessarily represent the official views of the granting agencies. The author has no conflicts of interest to report.

Contents

Contributors

MICKIE BHATIA • *Stem Cell and Cancer Research Institute, McMaster University, Hamilton, ON, Canada*

JUSTIN D. BLETHROW • *Thermo Fisher Scientific, San Jose, CA, USA*

LAURENCE M. BRILL • *Advanced Proteomics, Sanford|Burnham Medical Research Institute, La Jolla, CA, USA*

SEBASTIAN CAROTTA • *Molecular Immunology Division and Molecular Medicine Division, The Walter and Eliza Hall Institute of Medical Research, Parkville, Australia*

ANDREW M. CRAIN • *Stem Cell Biology and Regenerative Medicine Program, Sanford-Burnham Medical Research Institute, La Jolla, CA, Australia*

SATYABRATA DAS • *Division of Immunology, University of Iowa, Iowa City, IA, USA*

CASIMIR DE RHAM • *Immunology and Transplant Unit, Division of Immunology and Allergology, Geneva University Hospital and Medical School, Geneva, Switzerland; Division of Laboratory Medicine, Geneva University Hospital and Medical School, Geneva, Switzerland*

MICHA DRUKKER • *Institute of Stem Cell Biology and Regenerative Medicine, Stanford University School of Medicine, Stanford, CA, USA*

PAUL J. FAIRCHILD • *Sir William Dunn School of Pathology, University of Oxford, Oxford, UK*

MICHAEL HAYES • *Zavazava Lab, VAMC Iowa City, Iowa City, IA, USA*

ROXANNE HOLMES • *Sunnybrook Research Institute, Department of Immunology, University of Toronto, Toronto, ON, Canada*

JUNJIE HOU • *Advanced Proteomics, Sanford|Burnham Medical Research Institute, La Jolla, CA, USA*

NAOKI ICHIRYU • *Sir William Dunn School of Pathology, University of Oxford, Oxford, UK*

DAN S. KAUFMAN • *Department of Medicine and Stem Cell Institute, University of Minnesota, Minneapolis, MN, USA*

EUN-MI KIM • *Department of Internal Medicine, Division of Immunology, University of Iowa, Iowa City, IA, USA*

HANNES KLUMP • *Institute for Transfusion, University Hospital Essen, Essen, Germany*

DAVID A. KNORR • *Department of Medicine and Stem Cell Institute, University of Minnesota, Minneapolis, MN, USA*

JULIANE LADHOFF • *Berlin–Brandenburg Center for Regenerative Therapies (BCRT), Charité, Berlin, Germany; Laboratory of Molecular PsychiatryUniversitätsmedizin Charité, Berlin, Germany*

DANA LEVASSEUR • *Division of Immunology, University of Iowa, Iowa City, IA, USA*

HAYDN C.-Y. LIANG • *Department of Immunology, Sunnybrook Research Institute, University of Toronto, Toronto, ON, Canada*

FREDERICK LO • *Muscle Development and Regeneration Program, Sanford|Burnham Medical Research Institute, La Jolla, CA, USA*

GOHAR MANZAR • *Division of Immunology, Department of Internal Medicine and VAMC Iowa City, University of Iowa, Iowa City, IA, USA*

ZHENYA NI • *Department of Medicine and Stem Cell Institute, University of Minnesota, Minneapolis, MN, USA*

SANDRA PILAT • *Department of Medical Biology, The Walter and Eliza Hall Institute for Medical Research, Parkville, Victoria, Australia*

SUDHANSHU P. RAIKWAR • *Department of Internal Medicine, Division of Immunology, Roy J. and Lucille A. Carver College of Medicine, University of Iowa, Iowa City, IA, USA; Veterans Affairs Medical Center, Iowa City, IA, USA*

SHRAVANTI RAMPALII • *Stem Cell and Cancer Research Institute, McMaster University, Hamilton, ON, Canada*

ANGELIQUE SCHNERCH • *Stem Cell and Cancer Research Institute, McMaster University, Hamilton, ON, Canada*

MARTINA SEIFERT • *Institute of Medical Immunology and Berlin-Brandenburg Center for Regenerative Therapies (BCRT), Universitätsmedizin Charité, Berlin, Germany*

ILYAS SINGEC • *Stem Cell Biology and Regenerative Medicine Program, Sanford|Burnham Medical Research Institute, La Jolla, CA, USA*

EVAN Y. SNYDER • *Stem Cell Biology and Regenerative Medicine Program, Sanford|Burnham Medical Research Institute, La Jolla, CA, USA*

CHAD TANG • *Institute of Stem Cell Biology and Regenerative Medicine, Stanford University School of Medicine, Stanford, CA, USA*

BRIAN T.D. TOBE • *Stem Cell Biology and Regenerative Medicine Program, Sanford|Burnham Medical Research Institute, La Jolla, CA, USA*

JEAN VILLARD • *Immunology and Transplant Unit, Division of Immunology and Allergology, Geneva University Hospital and Medical School, Geneva, Switzerland; Division of Laboratory Medicine, Geneva University Hospital and Medical School, Geneva, Switzerland*

IRVING L. WEISSMAN • *Institute of Stem Cell Biology and Regenerative Medicine, Stanford University School of Medicine, Stanford, CA, USA*

DIETER A. WOLF • *Cancer Center, Sanford|Burnham Medical Research Institute, La Jolla, CA, USA*

NICHOLAS ZAVAZAVA • *Department of Internal Medicine and VAMC Iowa City, Division of Immunology, Roy J. and Lucille A. Carver College of Medicine, University of Iowa, Iowa City, IA, USA; Immunology Graduate Program, University of Iowa, Iowa City, IA, USA; Veterans Affairs Medical Center, Iowa City, IA, USA*

JUAN CARLOS ZÚÑIGA-PFLÜCKER • *Department of Immunology, Sunnybrook Research Institute, University of Toronto, Toronto, ON, Canada*

Chapter 1

Immune Privilege of Stem Cells

Naoki Ichiryu and Paul J. Fairchild

Abstract

Immune privilege provides protection to vital tissues or cells of the body when foreign antigens are introduced into these sites. The modern concept of relative immune privilege applies to a variety of tissues and anatomical structures, including the hair follicles and mucosal surfaces. Even sites of chronic inflammation and developing tumors may acquire immune privilege by recruiting immunoregulatory effector cells. Adult stem cells are no exception. For their importance and vitality, many adult stem cell populations are believed to be immune privileged. A preimplantation-stage embryo that derives from a totipotent stem cell (i.e., a fertilized oocyte) must be protected from maternal allo-rejection for successful implantation and development to occur. Embryonic stem cells, laboratory-derived cell lines of preimplantation blastocyst-origin, may, therefore, retain some of the immunological properties of the developing embryo. However, embryonic stem cells and their differentiated tissue derivatives transplanted into a recipient do not necessarily have an ability to subvert immune responses to the extent required to exploit their pluripotency for regenerative medicine. In this review, an extended definition of immune privilege is developed and the capacity of adult and embryonic stem cells to display both relative and acquired immune privilege is discussed. Furthermore, we explore how these intrinsic properties of stem cells may one day be harnessed for therapeutic gain.

Key words Embryonic stem cells, Induced pluripotent stem cells, T cells, Neuronal cells, Insulin-producing cells, Hematopoietic cells, Transplantation tolerance

Abbreviations

APC	Antigen-presenting cell
CCR4	CC chemokine receptor 4
DC	Dendritic cell
ES	Embryonic stem
HSC	Hematopoietic stem cell
IDO	Indoleamine 2,3 dioxygenase
IFN	Interferon
IL-10	Interleukin-10
IVD	Intervertebral disk
KIR	Killer cell immunoglobulin-like receptor
mCRPs	Membrane complement regulatory proteins
MHC	Major histocompatibility complex

Nicholas Zavazava (ed.), *Embryonic Stem Cell Immunobiology: Methods and Protocols*, Methods in Molecular Biology, vol. 1029, DOI 10.1007/978-1-62703-478-4_1, © Springer Science+Business Media New York 2013

MICA	MHC class I chain-related gene A
MIF	Migration inhibitory factor
MSC	Mesenchymal stem cell
NK	Natural killer cell
NSC	Neural stem cell
TGF-β	Transforming growth factor β
T_{reg}	Regulatory T cell
uNK	Uterine natural killer cell

1 Introduction

Immune privilege provides protection against immune-mediated injury to a variety of tissues in the body [1]. It was originally defined as a property of sites where an allogeneic transplant (typically a skin graft) could survive long term or indefinitely without being rejected by the recipient's immune system, even though the same graft would be rejected elsewhere in the body [2, 3]. Classic examples of immune-privileged sites are the brain and anterior chamber of the eye [4]. Maternal-fetal tolerance, or pregnancy-associated immune privilege, has also been similarly defined, since an embryo implanted in a mother's uterus represents a semi-allogeneic "graft" expressing paternal antigens. It is also established that not only sites, in terms of locations in the body, but also tissues themselves can be privileged [3]. This can be demonstrated by the survival of these tissues when transplanted ectopically and allogeneically (i.e., across immunological barriers) into nonimmune-privileged sites.

The modern definition of immune privilege has become extensive, applying to situations beyond transplantation and pregnancy. Various sources of evidence suggest that many tissues in the body can be placed on a spectrum of "relative immune privilege," each tissue or cell type assigned a certain degree of protection and tolerogenicity, albeit sometimes transient [5]. Relative immune privilege is conferred on a tissue as a result of multiple cellular and molecular events, often a subset of the properties displayed by classical immune-privileged sites. According to this definition, immune privilege is relative, not absolute. Consequently, it can fail or break down under some circumstances [5]. Additionally, the concept of "acquired immune privilege" challenges the idea that immune privilege is a hardwired characteristic of certain tissues. Acquired immune privilege is mediated by immunosuppressive effector cells, in particular $CD4^{+}FoxP3^{+}$ regulatory T cells (T_{reg}), which can induce local tolerance [6]. Since acquired immune privilege is not restricted to a certain anatomical location or structural feature of an organ, it can be induced anywhere in the body, among allografts, as well as tumors and infected tissues [7].

These new concepts of immune privilege have encouraged many researchers to seek opportunities for immune intervention and therapeutic manipulation for the successful treatment of disease. On the one hand, immune privilege acquired by tumor tissue is an obstacle to successful cancer immunotherapy. Therefore, a method to specifically abrogate the status of immune privilege from the tumor mass and the cancer stem cells that sustain it may greatly improve the efficiency of cancer immunotherapy. On the other hand, an immune privileged status may be conferred on an organ or a tissue allograft, in order to achieve the ultimate goal of transplantation, namely, the long-term acceptance of grafts with as little general immunosuppression as possible. Such an approach may be extended to the treatment of other conditions, such as preeclampsia and type 1 diabetes.

In the field of regenerative medicine, adult and embryonic stem (ES) cells potentially provide a sustainable supply of tissues for transplantation. However, since donor adult stem cells or ES cell-derived tissue will not necessarily be genetically identical to recipients, they remain under the threat of rejection. Therefore, the field of regenerative medicine is seeking solutions to the immunological barriers by manipulating immune privilege. Adult stem cells that populate most tissues in the body are vital for growth and homeostasis, and may be given special privileges by the immune system in order to be kept quiescent and uncompromised by deleterious autoimmune or inflammatory responses. Therefore, it is tempting to speculate that adult stem cells form a part of the immune-privileged compartment. In contrast, it is not clear to what extent ES cells display intrinsic and/or acquired immune privilege in vivo, given their artifactual nature when maintained in vitro. In this review, we address these issues by discussing evidence for intrinsic and acquired immune privilege among distinct populations of stem cells and how these may be exploited therapeutically.

2 Mechanisms of Immune Privilege

Immune privilege, as the name suggests, is a special benefit that is conferred upon a tissue by the immune system. The main characteristic of immune privilege is that it can subdue a competent immune system capable of mounting a fully fledged immune response (Fig. 1). There are two ways in which a potentially immunogenic tissue can avoid pursuit by the immune system: a passive form of antigenic invisibility or active immune regulation. Antigenic invisibility may be achieved by the lack of antigen presentation or by actively preventing its recognition. In this case, the host immune system is ignorant of the target tissue or its antigens. However, this mode of evasion is rather passive: although, in many cases, it may enhance the protection of immune-privileged sites, it

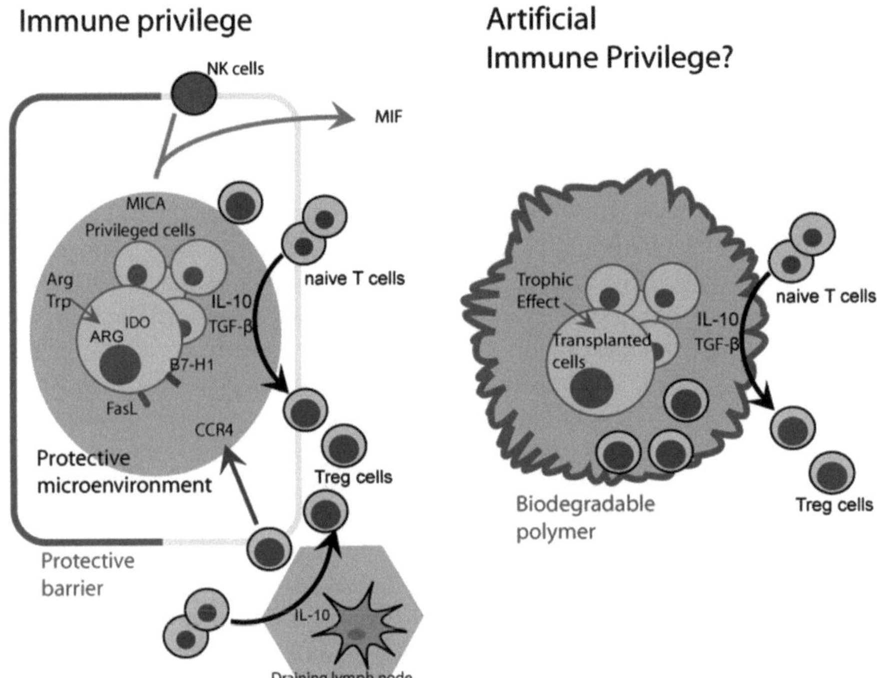

Fig. 1 Schematic diagrams of how immune privilege is achieved in intrinsic or acquired immune-privileged sites (*left*) and how immune privilege could be achieved artificially for transplanted cells (possibly derived from stem cells) (*right*) exploiting the mechanisms of immune privilege as well as providing a suitable environment with biodegradable polymers. Immune privilege found in certain tissues or organs is achieved as a result of multiple mechanisms that work in concert. The tissue is normally surrounded by, or "hidden" behind, a protective epithelial layer, such as the blood–brain barrier, to prevent or limit the access of professional immune effector cells. Within the tissue, there is an anti-inflammatory microenvironment, with altered local amino acid concentrations, soluble anti-inflammatory molecules, and surface molecules that antagonize inflammatory mediators. Even in the sites lacking the protective barrier, T_{reg} cells can invade the tissue and actively prevent T effector cells from activation, and also may polarize them to a regulatory phenotype by the action of TGF-β. IL-10-producing dendritic cells in a nearby draining lymph node can also signal naïve T cells to differentiate into regulatory cells in some forms of acquired immune privilege found in cancer. To recreate similar conditions in an artificial setting, biodegradable polymers or similar materials can be employed to serve as a protective barrier, as well as to retain some of the anti-inflammatory factors and create a microenvironment similar to that found in an immune-privileged tissue. T_{reg} cells that home to these tissues can induce acquired immune privilege, and actively promote tolerance rather than rejection

is neither essential nor sufficient for the establishment of immune privilege. Consequently, privilege achieved by this mechanism alone can be compromised when the immune system is primed against these antigens elsewhere in the body. On the other hand, active immune regulation can locally suppress a primed adaptive immune response against the target antigen. In this case the immune system is fully aware of the presence of this antigen-bearing tissue but is unable to reject it due to mechanisms that actively suppress the immune response. Most of the classical immune-privileged sites possess both of these mechanisms, while other tissues and sites may only display the latter.

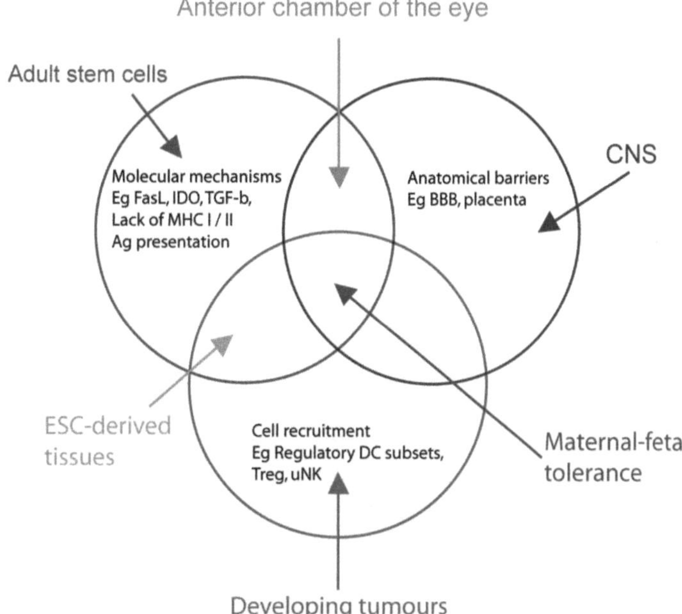

Fig. 2 The mechanisms of immune privilege are diverse, ranging from reliance on anatomical barriers to the deployment of a varied molecular arsenal and the recruitment of cell types capable of suppressing the immune response. These strategies may be used alone or in concert by individual cell types and complex tissues to secure their immunological integrity

2.1 Molecular Mechanisms

Immune privilege can be discussed at multiple levels (Fig. 2). At one end of the spectrum are small molecules (or lack of them) that play important roles in the induction and maintenance of immune privilege. Amino acids, the building blocks of proteins in an organism, are circulated and distributed among tissues. Essential amino acids, which cannot be synthesized within the body, need to be consumed as nutrients, and may limit protein synthesis. The consumption of essential amino acids is higher in rapidly growing cells, since DNA replication and cell division require continual synthesis of proteins. Lymphocytes are especially sensitive to amino acid starvation, since their function depends on rapid proliferation [8]. Therefore, limiting the availability of essential amino acids may interfere with the function of the immune system [9]. Consequently, lowering the local availability of some or all essential amino acids in a tissue may allow it to acquire immune privilege, and this mechanism is employed by a number of immune-privileged tissues.

Indoleamine 2,3-dioxygenase (IDO) is an intracellular enzyme that catabolizes the essential amino acid tryptophan. It was first described as an important factor in terms of immune privilege in pregnancy, because the inhibition of its enzymatic activity leads to a failure in the implantation of allogeneic, but not syngeneic, embryos in the mouse [10]. IDO has been explored extensively in its role in the induction of tolerance and maintenance of immune privilege. Expression of IDO, not only by the trophoblasts during pregnancy but also by subsets of dendritic cells (DCs), serves an important role in polarizing the adaptive immune response. Antigen presentation in the context of tryptophan deprivation not only disables the effector functions of lymphocytes but also polarizes $CD4^+$ T cells towards a T_{reg} phenotype. More recently, arginine catabolism by NOD and arginase 1 and 2 enzymes has also been implicated in the establishment of immune privilege [11]. Furthermore, it seems that this effect is not restricted to tryptophan and arginine, but is more broadly applied to many of the essential amino acids that modulate T lymphocyte responses via the mTOR pathway when depleted [12].

At the protein level, antigen presentation by the tissue via major histocompatibility complex (MHC) molecules is often modified in immune-privileged tissues. MHC antigen expression is an important mediator of the adaptive immune response, and is essential for the recognition of both self- and foreign antigens. The down-regulation of MHC I and II antigen presentation is a common feature shared by many immune-privileged sites. In the eye, for example, MHC I antigen presentation, which is ubiquitous in all nucleated cells in the body, is low or absent from many of the cell types, such as corneal endothelial cells. Similarly, MHC II expression is almost completely absent in the eye [3]. Similarly, trophoblast cells in human placenta display a unique pattern of MHC antigen expression. Trophoblast cells, derived from the implanted embryo, invade and integrate with the maternal tissue and blood supply, thereby serving as a marginal layer that separates the maternal immune system from the fetus [13]. Trophoblast cells carry paternal genes and therefore may be recognized by the maternal immune system as foreign. However, labyrinthine trophoblast cells, in direct contact with the maternal blood, fail to express either of the two classes of MHC molecules [13]. On the other hand, spongiotrophoblast cells that are in direct contact with the maternal decidua tissue do not express the conventional MHC class I molecules (HLA-A and HLA-B) but the expression of unconventional MHC class I molecules (HLA-C, HLA-G, HLA-E) is preserved [13].

The absence of conventional antigen presentation by MHC class I molecules is effective in evading immunosurveillance by the adaptive immune system. However, cells that lack MHC class I expression are susceptible to the natural killer (NK) cell response. NK cells are thought to operate in such a way that the level of killer-cell immunoglobulin-like receptors (KIR) on NK cells bound to MHC class I molecules of the target cells determine their cytotoxic activity. Therefore, in addition to "hiding" from antigenic recognition by the adaptive immune response, many of the immune-privileged tissues actively suppress NK cell responses. During pregnancy, a large accumulation of specialized NK cells is present in the uterus both in humans and in mice [14]. These uterine natural killer (uNK) cells (CD56^{bright}/CD16^{-}) differ from the conventional NK cells (CD56^{dim}/CD16^{+}) in their cytotoxic activities and other functions. uNK cells are thought to play an important role in regulating the trophoblast invasion into the decidua, and in tissue remodeling. In humans, while HLA-G and HLA-E expressed by some trophoblast cells are relatively non-polymorphic [15], HLA-C is polymorphic and may lead to a cytotoxic response by conventional NK or uNK cells. It is believed that the interaction between HLA-C and KIR is an important factor in successful pregnancy, and thus may contribute to population-specific skewing of HLA-C–KIR genotype combinations [13]. HLA-G and HLA-E have affinity to NK cell inhibitory receptors, preventing NK cell-induced apoptosis. However, it is unclear how some trophoblast cells that are in direct contact with the maternal blood circulation protect themselves against circulating NK cells.

Human hair follicles, sites of relative immune privilege with little MHC class I expression, were found to actively suppress NK cell responses by expression of macrophage migration inhibitory factor (MIF; an NK cell inhibitor) and low expression of MHC class I chain-related A (MICA) gene (an activator of NK cells). The relative expression of these genes seems to have an important role in the maintenance of immune privilege in hair follicles, since low expression of MIF and high expression of MICA are seen in hair follicles isolated from alopecia areata lesions, along with the accumulation of a large number of activated NK cells [16].

Surface expression of regulatory factors, such as Fas-ligand (FasL) and B7-H1 (programmed death-1 receptor, CD274), is implicated in the induction and maintenance of immune privilege [17]. B7-H1 is constitutively expressed in corneal endothelium and stromal cells, even when transplanted as an allograft. Against a background of 50 % spontaneous survival of corneal allografts transplanted between two strains of mice, the blockade of B7-H1 or its receptor PD-1 (expressed by CD4^{+} T cells) significantly reduced the rate of corneal allograft survival [18]. Many human tumors are also known to express B7-H1 which provides an advantage for tumor mass survival by causing infiltrating CD4^{+} PD-1^{+} T cells to undergo apoptosis [19]. Another pro-apoptotic factor, FasL, is also constitutively expressed in immune-privileged sites, including parts of the testis and the eye. However, an immunomodulatory effect and its role in the maintenance of immune privilege are complex [20]. The transgenic expression of FasL in some allografts was found to protect the tissue by inducing T cell apoptosis, but other studies have shown accelerated rejection of such grafts. This may be due to dose-dependency of FasL responses. FasL is known as a chemotactic factor for neutrophils and as an apoptotic factor for T cells, and it can exist in either membrane-bound or soluble forms [20].

Additionally, normal tissue and cells are protected from complement-induced lysis by expressing membrane complement regulatory proteins (mCRPs) [21]. It is believed that inhibition of complement-induced lysis is a general protective mechanism of host cells, but there are indications that these proteins are differentially expressed among tissues, depending on their function. CD49, one of the mCRPs, is strongly expressed in the genital tract, and is important for fertility. While human hepatocytes express a number of mCRPs and are, therefore, protected from complement attack [22], cancer cells may exploit this mechanism to inhibit complement-induced lysis by upregulating mCRP expression.

2.2 Micro-environment and Structural Features

Immune-privileged cells often express immunoregulatory and anti-inflammatory factors, such as interleukin-10 (IL-10) and transforming growth factor-β (TGF-β). These factors may exert their effects at a systemic level, but they are often retained in the tissue and therefore act locally. Therefore, anti-inflammatory factors released by a tissue can create a specialized microenvironment where the risk of tissue damage due to inflammation is minimized [23]. The depletion of essential amino acids, as discussed above, may also act against proinflammatory signals and locally inhibit T cell activation, at the same time polarizing infiltrating T cells towards conventional $CD4^+Foxp3^+$ T_{reg} cells, or other regulatory phenotypes.

At a tissue/organ level, immune-privileged cells are often surrounded by a protective cell layer that provides a physical barrier, well-known examples being the blood–brain barrier and the trophoblast layer in the placenta [4]. Only limited or regulated access by professional immune cells is allowed beyond the barrier. Antigen presentation may also be hindered by the lack of afferent lymphatics, while an absence of vascularization is an important factor in maintaining immune privilege in the anterior chamber of the eye [2]. Local immune privilege may also be affected by systemic changes. Immune privilege is known to be adjusted and modified according to age and biological time frame of an animal's lifespan by hormonal changes. For example, during pregnancy, the maternal immune system is strongly biased towards Th2 responses. In addition, the integrity of the immune system and its regulation may be affected by neurological signals, since high stress levels are known to be associated with increased failure of immune privilege [24].

2.3 Acquired Immune Privilege

Examples of acquired immune privilege are mainly found in induced local tolerance following allogeneic transplantation, in sites of chronic inflammation, and in the escape of tumors from immune surveillance [6]. These "abnormal" target cells usually present antigens which can be recognized and become a target for an immunological response. However, when tissues successfully acquire immune privilege, rejection is actively prevented by $CD4^+$ T_{reg} cells. T_{reg} cells can infiltrate the target tissue and provide local protection by producing immunoregulatory factors, such as TGF-β, at the same time inducing naïve peripheral T cells to acquire a regulatory phenotype. During pregnancy, in addition to a Th2 bias, a larger proportion of T cells becomes committed to a $CD4^{+}FoxP3^{+}$ regulatory phenotype, both in humans and mice [25, 26]. These T_{reg} cells infiltrate the placenta and confer extra protection on a semi-allogeneic conceptus. In addition to T_{reg} cells inducing differentiation of naïve T cells to T_{reg} cells, through expression of CTLA-4 and TGF-β, some specialized populations of DCs can also mediate the induction of T_{reg} cells. A subset of DCs found in proximity to immune-privileged tissue do not express high levels of co-stimulatory factors but express instead IDO, as well as the anti-inflammatory cytokines IL-10 and TGF-β.

Many developing tumors also acquire immune privilege. Hodgkin's lymphoma-derived cells produce ligands for CC chemokine receptor 4 (CCR4) which is expressed by $CD4^{+}$ cells with anti-proliferative effector action, conferring on the tumor cells an acquired immune privilege [27]. Ovarian carcinoma was found to acquire immune privilege by the presence of IL-10-producing DCs in the draining lymph nodes [28]. The presence of chronic inflammation may also be an important factor. A mouse model of skin cancer development, involving mutagenesis followed by a chemical treatment to mimic chronic inflammation, allowed the outgrowth of malignant cells in the form of a papilloma [29].

3 Immune Privilege of Stem Cells

Adult stem cells are absolutely required for the growth, homeostasis, and rejuvenation of tissues. They have multiple but restricted capacity to differentiate into the cell types that make up the tissue in which they reside. Since these stem cells are typically long-lived and vital for the routine turnover of tissues and their response to injury, it is essential that they are protected from adverse autoimmune responses or local damage from chronic inflammation. Furthermore, since stem cell pools typically persist for the entire life span of an organism, they are maintained quiescent in their specialized niches until their proliferation and differentiation are required. Hematopoietic stem cells (HSCs), for example, are maintained in a quiescent state by their surrounding microenvironment as well as by intrinsic transcriptional regulators [30, 31]. It is possible that these adult stem cell niches serve as a protective microenvironment in addition to stem cells themselves being immune privileged. Embryonic stem cells, with a wider range of differentiation capacity, have no in vivo counterpart in an adult organism, and hence have no niche to which they could naturally home when transplanted into an adult organism. Nevertheless, since they are derived from the inner cell mass of preimplantation-stage blastocysts, they may resemble the early embryo in terms of their intrinsic capacity for immune privilege [32]. Thus, a common feature shared by both adult and embryonic stem cells is that they are very likely immune privileged in their native location, since immune privilege has evolved to protect vital organs for growth, survival, and reproduction [3].

3.1 Immune Privilege of Adult Stem Cells

Various adult stem cell populations have been identified and some have been well characterized in the context of immune privilege. Mesenchymal stem cells (MSCs) are well known for their capacity to suppress inflammation and inhibit the immune response [33]. MSCs and the surrounding stromal cells that form MSC niches inhibit cyclin-D2 expression which disrupts the cell cycle of a variety of cell types, including T cells [34, 35]. B7-H1, expressed by MSCs in response to increased interferon-γ (IFN-γ) production by T cells, can, in turn, down-modulate the effector functions of activated T cells through PD-1 ligation [36]. The immunosuppressive capacity of MSCs has been applied to many different settings from tissue repair to the prevention of graft-versus-host disease [37]. Hiyama et al. recently succeeded in blocking degeneration of intervertebral discs (IVDs) in a surgically induced canine nucleotomy model by transplanting MSCs. MSCs contributed to the maintenance of immune privilege in IVDs by producing FasL [38].

Neural stem cells (NSC) are immune privileged, not only in terms of their native niche but also in terms of allogeneic transplantation into a non-privileged site. A study by Hori et al. demonstrated that neural progenitor cells, a group of cells which contain NSC as well as more differentiated neural precursor cells, were unable to sensitize the allogeneic recipient when transplanted beneath the kidney capsule in the form of a neurosphere, whereas neonatal cerebellum, transplanted to the same site, was rejected [39]. However, once an animal had been sensitized to the alloantigens post transplantation, the graft was rejected, suggesting that NSC passively evade immune surveillance; T cells were kept ignorant of the presence of antigen, instead of acquiring a local or a systemic antigen-specific tolerance.

3.2 Immune Privilege of ES Cells

Use of ES cell-derived tissues in cell replacement therapy is an attractive option for the purposes of regenerative medicine, since, unlike adult stem cells, ES cells could potentially be propagated and expanded indefinitely in vitro, to meet the growing demand. There are reasons to speculate that ES cell-derived tissues are less immunogenic compared to solid organ transplants or other tissues. ES cells lack or have little surface MHC class I expression, unless induced by IFN-γ. MHC class II expression is also absent, but, unlike MHC class I, is not inducible, greatly limiting the direct presentation of alloantigen to the recipient immune system [40]. Furthermore, in contrast to a solid organ allograft, ES cell-derived tissues do not carry donor antigen-presenting cells (APCs), including the DCs that initiate primary immune responses. This suggests that the direct presentation of alloantigen by donor DCs to recipient effector T cells, normally responsible for acute rejection of a graft, does not occur. However, once antigens from ES cell-derived tissues have been processed and presented by the recipient's own DCs, the immune system can recognize the allograft as foreign and reject it. Furthermore, undifferentiated ES cells would be susceptible to NK cell responses since they lack MHC class I expression. Whether ES cells and their derivatives actively suppress an NK cell response and other aspects of the maternal immune response is unclear.

Gene micro-array analysis of human ES cells by Grinnemo et al. showed that they fail to express classic immune-privileged factors like TGF-β, FasL, or IL-10 [40]. Contrary to these findings, however, earlier experiments suggested that undifferentiated human ES cells were either immunosuppressive or did not elicit rejection by mouse recipients [41, 42]. Nevertheless, a recent study with better cell-fate tracking of transplanted human ES cells showed that they are, in fact, rejected when injected into mouse muscle [43, 44]. This is not surprising since transplantation of ES cells into allogeneic or xenogeneic nonimmune-privileged sites exposes them to a harsh nonnative environment. Many reports suggest that ES cells and their derivatives have the potential to differentiate and contribute to tissue repair and regeneration, but such studies were invariably conducted in the absence of a competent immune system. For example, a study by Dai et al. showed that human ES cell-derived cardiomyocytes can successfully improve rat heart function following ischemic damage, but only when nude athymic recipients, rather than immune-competent rats, were used [45]. It appears, therefore, that, in spite of the potential immune privilege displayed by ES cell-derived tissues, allograft rejection remains a major obstacle to regenerative medicine, no less challenging than the rejection of solid organ allografts [46].

So why the controversy? One possibility is that, since it is difficult in many of these experiments to include appropriate controls, some of the apparent immune privilege of ES cells may not be ES cell specific. An attempt to induce better immune suppression and tolerance by co-transplantation of human ES cells and human MSCs into rat hearts suggested that ES cells alone did not result in an improvement in heart function following transplantation [47]. Nevertheless, when co-transplanted with MSCs, heart function was improved, although the effect was only modest. Authors of the study discuss that there may be a more significant benefit from using a larger number of transplanted cells as a result of their trophic effect on the recipient's own tissues, which is in line with the achievement of immune privilege through amino acid depletion discussed above. Dose-dependency of immunosuppression by ES cells has also been suggested [42].

Evidence suggests, therefore, that immune privilege of ES cells is, at best, relatively weak. In order to define the magnitude of the immunological barriers that could be tolerated using tissues differentiated from ES cells, our own laboratory derived a number of novel ES cell lines from strains of mice differing from the recipient strain at defined genetic loci [48]. Tissues were transplanted under the kidney capsule of recipients where they were able to form conventional teratomas. By changing the combination of ES cell and recipient mouse strains, the exact extent of the immunological disparity could be readily controlled and the propensity for rejection monitored as a function of infiltrating effector cells and the size and integrity of the teratoma. Using this model, it was found that ES cells could not overcome a full MHC mismatch [48] while in a related study, relative immune privilege of ES cells achieved by the absence of antigen presentation did not persist after terminal differentiation into insulin-producing cells due to an eventual up-regulation of MHC class I expression as the cells matured [49]. Even a mismatch at multiple minor histocompatibility antigens was sufficient for ES cell grafts to be rejected [48]. However, a relative immune privilege could be observed; immune intervention with either anti-CD4 or anti-CD8 nondepleting antibodies was sufficient to induce infectious tolerance through the recruitment of T_{reg} cells, whereas additional co-stimulatory-blockade was necessary for the acceptance of skin grafts [48]. In a model where the only mismatch between the recipient and the donor ES cell line was a single MHC class I antigen (H-2K^{b}), 50 % acceptance could be observed spontaneously in the absence of immune intervention [50].

3.3 Application of Immune Privilege to Regenerative Medicine

The culmination of extensive studies using animal models of disease led to the approval in January 2009 of the first clinical trials of human ES cell-derived tissue by Geron Corporation. This represents a great advance in the field of regenerative medicine, and the result of this trial will inevitably influence the future direction of the field. In addition to the primary trial, which aims to enhance recovery from spinal cord injuries with ES cell-derived oligodendrocyte progenitors, the eye may be the next target of cell replacement therapy. Both of these sites are realistic targets since they may benefit from the intrinsic mechanisms of immune privilege offered by their anatomical location. Nevertheless, despite these significant advances, the field still lacks a convenient model in which the immunogenicity of human ES cell-derived tissues may be tested against a competent human immune system. Currently, "humanized" trimera mouse models [51] are probably the best available. These recipients are generated from immunocompetent mice bearing intact lymphoid organs, which have been irradiated, injected with SCID bone marrow, and reconstituted with human peripheral blood mononuclear cells [52]. Human ES cell-derived tissues can be transplanted under the kidney capsule of such trimera mice in order to investigate their immunogenicity. However, this represents an extremely complex model which only partially mimics the human immune system.

Although numerous studies now indicate that the capacity for immune privilege may not be sufficient to secure the survival of ES cell-derived tissues in the context of cell replacement therapy, the prospect remains for immune privilege to act in concert with established protocols for the induction of tolerance to overcome the immunological barriers encountered in the clinic. Like some cases of successful HSC transplantation which resulted in the establishment of mixed chimerism and systemic donor-specific tolerance, stem cells derived from rat preimplantation-stage embryos were reported to differentiate along the hematopoietic lineage and persist in a recipient rat following their injection into the hepatic portal vein. This permitted the acceptance of subsequent grafts from the strain of rats from which the ES cell-like cells were derived, consistent with antigen-specific tolerance [53]. However, the achievement of mixed chimerism is difficult in humans, and this mode of tolerance, without the irradiation of the recipient, has not been achieved with either mouse or human undifferentiated ES cells. For example, Magliocca et al. assessed whether transplantation of undifferentiated mouse ES cells could condition the recipient mouse to accept a differentiated tissue derived from the same ES cell line. Although the primary undifferentiated inoculum was accepted, as demonstrated by teratoma formation, the differentiated graft was subsequently rejected [54]. Recently, a stable mixed chimerism and donor-specific tolerance in mice were achieved by Bonde et al. through the administration of hematopoietic cells differentiated in vitro from a mouse ES cell line to irradiated recipients [55]. Long-term graft survival was observed when these chimeric recipients subsequently received cardiac allografts from the donor mouse strain. Even though this work may suggest a potential method to induce antigen-specific tolerance using ES cell-derived tissues, it is uncertain whether it can be applied to humans as a result of the level of conditioning required to permit the acceptance of donor HSC. Furthermore, mixed chimerism is wholly dependent on the thymic function of the recipient, a parameter which is known to be significantly compromised in the elderly, who remain the most likely beneficiaries of such therapies. However, an alternative approach, which may prove more amenable to translation, is to combine the propensity of ES cell-derived grafts for immune privilege with the administration of nondepleting CD4 and CD8 monoclonal antibodies. Studies in mice have already shown how acceptance across a full MHC barrier may be achieved with minimal conditioning of the host, suggesting that it may ultimately be easier to gain acceptance of ES cell-derived tissues than allografts from any other source [48].

4 Conclusions

Immune privilege is not achieved by any one definitive factor, but rather the combination of various mechanisms that may act in concert (Fig. 2). Future approaches to manipulating immune privilege would, therefore, need to consider various factors at multiple levels. For example, a study by Tat et al. demonstrated that retinal progenitor cells can foster immune privilege when implanted with biodegradable polymers to which the cells and released molecules can attach (Fig. 1) [23]. This work suggests that, for cells or tissues to be privileged, they need to be in the correct, perhaps protective, microenvironment. Evidence to date supports the presence of a relative immune privilege for both adult and ES cells. However, it is also clear that their privileged status is not absolute and is, in reality, rather fragile outside their native niche. The successful application of adult stem cell transplantation and ES cell-derived tissues to regenerative medicine will need, therefore, to consider ways of exploiting the intrinsic privileged nature of these cells while actively recruiting the regulatory capacity of the immune system for the induction of acquired immune privilege.

Acknowledgments

We are grateful to Dr Stephen Cobbold for helpful discussions. Work on immune privilege in the authors' laboratory is supported by seed funding from the Oxford Stem Cell Institute and by grant G0802538 from the Medical Research Council (UK).

References

1. Wahl SM, Wen J, Moutsopoulos N (2006) TGF-β: a mobile purveyor of immune privilege. Immunol Rev 213:213–227
2. Simpson E (2006) A historical perspective on immunological privilege. Immunol Rev 213: 12–22
3. Streilein JW (2003) Ocular immune privilege: therapeutic opportunities from an experiment of nature. Nat Rev Immunol 3:879–889
4. Niederkorn JY (2006) See no evil, hear no evil, do no evil: the lessons of immune privilege. Nat Immunol 7:354–359
5. Paus R, Nickoloff BJ, Ito T (2005) A "hairy" privilege. Trends Immunol 26:32–40
6. Mellor AL, Munn DH (2008) Creating immune privilege: active local suppression that benefits friends, but protects foes. Nat Rev Immunol 8:74–80
7. Cobbold SP, Adams E, Graca L et al (2006) Immune privilege induced by regulatory T cells in transplantation tolerance. Immunol Rev 213:239–255
8. Chuang JC, Yu CL, Wang SR (1990) Modulation of human lymphocyte proliferation by amino acids. Clin Exp Immunol 81:173–176
9. Calder PC (2006) Branched-chain aminoacids and immunity. J Nutr 136:288S–293S
10. Munn DH, Zhou M, Attwood JT et al (1998) Prevention of allogeneic fetal rejection by tryptophan catabolism. Science 281: 1191–1193
11. Bronte V, Zanovello P (2005) Regulation of immune responses by L-arginine metabolism. Nat Rev Immunol 5:641–654
12. Cobbold SP, Adams E, Farquhar CA et al (2009) Infectious tolerance via the consump-

tion of essential amino acids and mTOR signaling. Proc Natl Acad Sci 106:12055–12060
13. Moffett A, Loke C (2006) Immunology of placentation in eutherian mammals. Nat Rev Immunol 6:584–594
14. Riley JK, Yokoyama WM (2008) NK cell tolerance and the maternal-fetal interface. Am J Reprod Immunol 59:371–387
15. Hunt JS, Petroff MG, McIntire RH et al (2005) HLA-G and immune tolerance in pregnancy. FASEB J 19:681–693
16. Ito T, Ito N, Saatoff M et al (2008) Maintenance of hair follicle immune privilege is linked to prevention of NK cell attack. J Invest Dermatol 128:1196–1206
17. Keir ME, Butte MJ, Freeman GJ et al (2008) PD-1 and its ligands in tolerance and immunity. Annu Rev Immunol 26:677–704
18. Hori J, Wang M, Miyashita M et al (2006) B7-H1-induced apoptosis as a mechanism of immune privilege of corneal allografts. J Immunol 177:5928–5935
19. Dong H, Strome SE, Salomao DR et al (2002) Tumor-associated B7-H1 promotes T-cell apoptosis: a potential mechanism of immune evasion. Nat Med 8:793–800
20. Askenasy N, Yolcu ES, Yaniv I et al (2005) Induction of tolerance using Fas ligand: a double-edged immunomodulator. Blood 105:1396–1404
21. Fishelson Z, Donin N, Zell S et al (2003) Obstacles to cancer immunotherapy: expression of membrane complement regulatory proteins (mCRPs) in tumors. Mol Immunol 40:109–123
22. Halme J, Sachse M, Vogel H et al (2009) Primary human hepatocytes are protected against complement by multiple regulators. Mol Immunol 46:2284–2289
23. Tat FN, Lavik E, Keino H et al (2007) Creating an immune-privileged site using retinal progenitor cells and biodegradable polymers. Stem Cells 25:1552–1559
24. Arck PC, Gilhar A, Bienenstock J et al (2008) The alchemy of immune privilege explored from a neuroimmunological perspective. Curr Opin Pharm 8:480–489
25. Somerset DA, Zheng Y, Kilby MD et al (2004) Normal human pregnancy is associated with an elevation in the immune suppressive CD25$^+$CD4$^+$ regulatory T-cell subset. Immunology 112:38–43
26. Aluvihare VR, Kallikourdis M, Betz AG (2004) Regulatory T cells mediate maternal tolerance to the fetus. Nat Immunol 5:266–271
27. Ishida T, Ishii T, Inagaki A et al (2006) Specific recruitment of CC chemokine receptor 4-positive regulatory T cells in Hodgkin lymphoma fosters immune privilege. Cancer Res 66:5716–5722
28. Curiel TJ, Coukos G, Zou L et al (2004) Specific recruitment of regulatory T cells in ovarian carcinoma fosters immune privilege and predicts reduced survival. Nat Med 10:942–949
29. Muller AJ, Sharma MD, Chandler PR et al (2008) Chronic inflammation that facilitates tumor progression creates local immune suppression by inducing indoleamine 2,3 dioxygenase. Proc Natl Acad Sci U S A 105: 17073–17078
30. Liu Y, Elf SE, Miyata Y et al (2009) p53 regulates hematopoietic stem cell quiescence. Cell Stem Cell 4:37–48
31. Wilson A, Trumpp A (2006) Bone-marrow haematopoietic-stem-cell niches. Nat Rev Immunol 6:93–106
32. Fändrich F, Dresske B, Bader M et al (2002) Embryonic stem cells share immune-privileged features relevant for tolerance induction. J Mol Med 80:343–350
33. Uccelli A, Moretta L, Pistoia V (2008) Mesenchymal stem cells in health and disease. Nat Rev Immunol 8:726–736
34. Jones S, Horwood N, Cope A et al (2007) The antiproliferative effect of mesenchymal stem cells is a fundamental property shared by all stromal cells. J Immunol 179:2824–2831
35. Ramasamy R, Lam EWF, Soeiro I et al (2007) Mesenchymal stem cells inhibit proliferation and apoptosis of tumor cells: impact on in vivo tumor growth. Leukemia 21:304–310
36. Sheng H, Wang Y, Jin Y et al (2008) A critical role of IFN? in priming MSC-mediated suppression of T cell proliferation through upregulation of B7-H1. Cell Res 18:846–857
37. Francesco Dazzi Federica MM-B (2008) Mesenchymal stem cells for graft-versus-host disease: close encounters with T cells. Eur J Immunol 38:1479–1482
38. Hiyama A, Mochida J, Iwashina T et al (2008) Transplantation of mesenchymal stem cells in a canine disc degeneration model. J Orth Res 26:589–600
39. Hori J, Ng TF, Shatos M et al (2007) Neural progenitor cells lack immunogenicity and resist destruction as allografts. Ocul Immunol Inflamm 15:261–273
40. Grinnemo KH, Kumagai-Braesch M, MaÌŠnsson-Broberg A et al (2006) Human embryonic stem cells are immunogenic in allogeneic and xenogeneic settings. Reprod BioMed Online 13:712–724
41. Li L, Baroja ML, Majumdar A et al (2004) Human embryonic stem cells possess immune-privileged properties. Stem Cells 22:448–456

42. Koch CA, Geraldes P, Platt JL (2008) Immunosuppression by embryonic stem cells. Stem Cells 26:89–98
43. Chidgey AP, Boyd RL (2008) Immune privilege for stem cells: not as simple as it looked. Cell Stem Cell 3:357–358
44. Swijnenburg R-J, Schrepfer S, Govaert JA et al (2008) Immunosuppressive therapy mitigates immunological rejection of human embryonic stem cell xenografts. Proc Natl Acad Sci 105:12991–12996
45. Dai W, Field LJ, Rubart M et al (2007) Survival and maturation of human embryonic stem cell-derived cardiomyocytes in rat hearts. J Mol Cell Cardiol 43:504–516
46. Fairchild PJ, Cartland S, Nolan KF et al (2004) Embryonic stem cells and the challenge of transplantation tolerance. Trends Immunol 25:465–470
47. Puymirat E, Geha R, Tomescot A et al (2008) Can mesenchymal stem cells induce tolerance to cotransplanted human embryonic stem cells? Mol Ther 17:176–182
48. Robertson NJ, Brook FA, Gardner RL et al (2007) Embryonic stem cell-derived tissues are immunogenic but their inherent immune privilege promotes the induction of tolerance. Proc Natl Acad Sci U S A 104:20920–20925
49. Wu DC, Boyd AS, Wood KJ (2008) Embryonic stem cells and their differentiated derivatives have a fragile immune privilege but still represent novel targets of immune attack. Stem Cells 26:1939–1950
50. Lui KO, Waldmann H, Fairchild PJ (2009) Embryonic stem cells: overcoming the immunological barriers to cell replacement therapy. Curr Stem Cell Res Ther 4:70–80
51. Drukker M (2006) Immunogenicity of embryonic stem cells and their progeny. In: Lanza R, Klimanskaya I (eds) Methods enzymol. Academic, San Diego, CA
52. Lubin I, Segall H, Marcus H et al (1994) Engraftment of human peripheral blood lymphocytes in normal strains of mice. Blood 83:2368–2381
53. Fändrich F, Lin X, Chai GX et al (2002) Preimplantation-stage stem cells induce long-term allogeneic graft acceptance without supplementary host conditioning. Nat Med 8:171–178
54. Magliocca JF, Held IKA, Odorico JS (2006) Undifferentiated murine embryonic stem cells cannot induce portal tolerance but may possess immune privilege secondary to reduced major histocompatibility complex antigen expression. Stem Cells Dev 15:707–717
55. Bonde S, Chan KM, Zavazava N (2008) ES-cell derived hematopoietic cells induce transplantation tolerance. PLoS One 3: e3212

Chapter 2

Immunogenicity of In Vitro Maintained and Matured Populations: Potential Barriers to Engraftment of Human Pluripotent Stem Cell Derivatives

Chad Tang, Irving L. Weissman, and Micha Drukker

Abstract

The potential to develop into any cell type makes human pluripotent stem cells (hPSCs) one of the most promising sources for regenerative treatments. Hurdles to their clinical applications include (1) formation of heterogeneously differentiated cultures, (2) the risk of teratoma formation from residual undifferentiated cells, and (3) immune rejection of engrafted cells. The recent production of human isogenic (genetically identical) induced PSCs (hiPSCs) has been proposed as a "solution" to the histocompatibility barrier. In theory, differentiated cells derived from patient-specific hiPSC lines should be histocompatible to their donor/recipient. However, propagation, maintenance, and non-physiologic differentiation of hPSCs in vitro may produce other, likely less powerful, immune responses. In light of recent progress towards the clinical application of hPSCs, this review focuses on two antigen presentation phenomena that may lead to rejection of isogenic hPSC derivates: namely, the expression of aberrant antigens as a result of long-term in vitro maintenance conditions or incomplete somatic cell reprogramming, and the unbalanced presentation of receptors and ligands involved in immune recognition due to accelerated differentiation. Finally, we discuss immunosuppressive approaches that could potentially address these immunological concerns.

Key words Antigen presentation, Immune surveillance, Sialic acid, Xenoantigen, Episomal, Non-integrating, Teratomas

1 Introduction

Major and minor histocompatibility complex (MHC and mHC, respectively) antigens belong to a large and diverse group of molecules involved in immune recognition and graft rejection. Classical graft rejection responses stem primarily from structural differences between donor and host antigens, most prominently those belonging to the MHC family (reviewed in ref. 1). Recent experiments have demonstrated that differentiated human embryonic stem cells (hESCs) express MHC class I (MHC-I) molecules [2, 3]. As such, hESC derivatives are expected to promote allorejection responses similar to those observed following organ transplantation [4].

Nicholas Zavazava (ed.), *Embryonic Stem Cell Immunobiology: Methods and Protocols*, Methods in Molecular Biology, vol. 1029, DOI 10.1007/978-1-62703-478-4_2, © Springer Science+Business Media New York 2013

With advancements toward production of patient-specific hPSCs by parthenogenesis [5], somatic cell nuclear transfer of oocytes [6], and induction of pluripotency [7], rejection based on MHC mismatches may become technically avoidable. Here, we discuss experiments indicating that aberrant antigens and unbalanced presentation of immunologic signals that develop due to in vitro maintenance and differentiation may promote immune responses even against grafts derived from isogenic hPSCs. We primarily discuss immunologic hurdles relevant to hiPSC derivatives, since hiPSC lines may become a main source of patient-matched grafts. Discussions of immunologic considerations for allogeneic hPSC transplantation are covered elsewhere [8, 9].

We first discuss improper immune antigen presentation by hPSCs as a result of long-term maintenance in vitro. At least four possible sources and mechanisms are involved in incorporation and production of aberrant immune antigens, including (1) animal-derived [10, 11] and (2) non-physiologic media compounds [12] (i.e., high concentrations of hormones or antibiotics), (3) genetic abnormalities that lead to formation of atypical antigens [13, 14], and (4) incomplete reprogramming of somatic cells [15]. The second concern that we review here is rejection of hPSC progeny that are immunologically immature. Interactions of the immune system with somatic cells during fetal development and following birth gradually shape the presentation of activating and inhibiting signals required for immune surveillance [16]. Such fine-tuning of immune ligand presentation is unlikely to take place during the rapid course of hPSCs differentiation in vitro. These non-physiological conditions may produce somatic derivatives expressing residual embryonic antigens and/or exhibit an imbalanced repertoire of surface ligands necessary for immune cell inhibition.

Importantly, evidence highlight here indicate that both phenomena may lead to immune rejection and that isogenic and allogeneic hPSC derivatives are equally at risk. Still, it is likely that immune responses against aberrant antigens and immunologically immature cells will be more indolent than rejection processes targeting mismatched histocompatibility antigens expressed by allogeneic hPSC derivatives. We therefore also discuss milder immunosuppression regimes that could potentially attenuate these anticipated weaker rejection processes. Since the number of studies directly examining our hypothesized rejection is relatively small, we go on to propose experiments that should provide greater understanding of the immunological properties of hPSC-derived isogenic grafts.

1.1 Maintenance of hPSC May Result in Aberrant Antigen Presentation

The reliance on animal products for propagation of hPSCs raises the concern of xenoantigen incorporation (reviewed in ref. 17). For example, Martin et al. [10] showed that hESCs present the nonhuman sialic acid *N*-glycolylneuraminic (Neu5Gc) when co-cultured with mouse embryonic fibroblasts (MEFs) and animal serum. Uptake of Neu5Gc by human cells results in the substitution of *N*-acetylneuraminic acid (Neu5Ac), a glycan frequently added to human surface proteins [18]. Humans are unable to synthesize Neu5Gc [19]. However, most individuals develop glycan targeting antibodies as this glycan is introduced through animal and bacterial products [18]. This raises a concern for rejection of hPSC derivatives by anti-Neu5Gc antibodies as exposure of hESCs to human sera with such antibodies has resulted in complement deposition and cell death [10]. Notably, a follow-up study by Cerdan et al. [20] challenged these data by showing that complement fixation via Neu5Gc targeting antibodies did not result in significant hESC death, although the tested cell lines expressed Neu5Gc. Although these studies have been critical in establishing proof-of-concept for the risks associated with xenoantigen presentation by hPSCs, they were limited to direct complement-mediated lysis assays in vitro, and did not include exposure of hESCs to antibodies and immune cells in vivo. Therefore, conclusion of the immunological consequence of Neu5Gc incorporation awaits further analysis. Since it is likely that other xenoantigens are incorporated into hPSCs in vitro, we propose that ensuring transplantation safety requires broader xenoantigen characterization efforts.

To circumvent the possibility of xenoantigen incorporation, several laboratories have developed methods to derive and culture hPSCs without animal products [21–24]. Culture protocols utilizing human serum, defined medias, human feeder layers, and/or synthetic polymeric surfaces have been shown to be effective [22, 24]. However, none of these conditions have yet to be fully optimized or broadly adopted. Standard hPSC culture conditions, which include MEFs and animal sera, remain widely used and are considered better suited for long-term support of hPSC self-renewal [25].

Importantly, high concentration of media constituents may also alter the antigen signature of hPSCs and their derivatives. For example, CD30 is expressed by hESCs that are cultured in the presence of animal-free knock-out serum due to high ascorbate levels [12]. The concept that changes in culture conditions may alter antigen presentation by hPSCs has also been further supported by Newman and Cooper who reported that both hESCs and hiPSCs exhibit lab-specific gene expression signatures [26]. These results highlight the need to extensively characterize and optimize the effects of media formulations on antigen presentation by hPSCs as they may lead to changes in immunogenicity.

Genomic rearrangements of hPSCs have been described by a number of studies and may also lead to antigenic changes. Draper et al. [13] have characterized a gain of chromosome 12 and a 17q segment in H1, H7, and H12 hESC lines following long-term culture. Other groups have described additional chromosomal lesions in other cell lines [14, 27]. Cells bearing genetic abnormalities are likely selected in culture when the modifications provide survival and growth advantages [13, 14, 27]. Werbowetski-Oglivie et al. [28] have demonstrated this correlation by producing two genetically abnormal hESC sublines through prolonged culturing. These lines were found to exhibit an order of magnitude increase in the frequency of tumor-initiating cells and significantly higher proliferation capacity [28]. Genomic rearrangements in hPSCs have also been shown to correlate with aberrant surface antigen presentation. For example, Herszfeld et al. [29] discovered that karyotypically abnormal hESCs exhibit elevated CD30 levels. The amplification of the CD30 gene in hESCs indicates that this marker may provide a survival/growth advantage, as previously shown in Hodgkin's lymphoma cells [30]. Although a direct link between genetic abnormalities and increased immunogenicity has yet to be demonstrated, given the frequency and number of genetic lesions reported, we hypothesize that hPSC subclones may exhibit disparate immunological properties.

Reprogramming of somatic cells into hiPSCs may also results in aberrant antigen presentation due to partial transcriptional memory retained from the epigenetic signature of the original tissue [15]. These marks were shown to correlate with disparities in gene and miRNA transcription compared to hESCs even after prolonged culturing period [31–33], although a different study reported contrary results [34]. To test whether these transcriptional disparities increase the immunogenicity of iPSC Zhao et al. compared the survival of transplanted mouse isogenic iPSCs with isogenic ESCs. They found that the isogenic transplants derived from mouse iPSCs stimulate immune responses, while those derived from isogenic mouse ESCs did not [32]. Importantly, they confirmed that retroviral integration were not the cause of the heightened immunogenicity as iPSCs produced through episomal (non-integrative) plasmid exhibited the same level of immunogenicity. Epigenetic abnormalities have also been reported in hESCs that were derived from parthenogenetic tumors and in SCNT-derived ESCs and fetuses. For example, Stelzer et al. [33] reported that differentiated parthenogenetic hiPSCs exhibit altered trophectoderm, liver, and muscle gene transcription profiles. Animals produced through SCNT have also been found to exhibit faulty expression of developmental genes due to incomplete reprogramming (reviewed in ref. 35). Taken together, these data indicate that epigenetic alternations can lead to elevated immunogenicity of hPSCs.

We therefore propose to systematically compare hPSC derivatives with the equivalent somatic cells in vivo, to detect differences in surface antigen presentations. Functional assays should follow to evaluate the clinical consequences of antigenic discrepancies. Such characterization efforts are already underway including those coordinated through the International Stem Cell Initiative Consortium, which have determined preliminary census properties of hPSCs by characterizing dozens of hESC lines [36, 37], and culture [38] and storage conditions [39].

1.2 Transplantation of Non-matched hPSC Lines Elicits Alloreactive Immune Responses

To discuss potential rejection pathways of isogenic hPSC lines, we first introduce the known interactions of hPSC allografts with the immune system. Classical MHC-I molecules are ubiquitously expressed heterodimers consisting of a single human leukocyte antigen (HLA)-A, -B, or -C chain and a β2-microglobulin molecule that together present a short peptide sequence sampled from intracellular proteins. The interactions of T-cell receptors (TCRs) with MHC-I molecules are one of the fundamental mechanisms distinguishing peptides as self or nonself. In transplantation settings, a small fraction of the hosts' TCRs interact with donor's MHC-I molecules, leading to maturation of cytotoxic T-cells and consequently the development of immune responses [1].

The expression of MHC-I molecules is developmentally regulated, increasing from low levels on pluripotent cells to higher levels throughout fetal development [16, 40–43]. Multiple studies have shown that MHC-I molecules are similarly expressed at low levels on human [2, 3, 44–46] and mouse [2, 47, 48] PSCs and increase on differentiated derivatives, albeit ultimate expression levels are lower than somatic cells. Although MHC-I presentation by ESC-derivatives is lower than somatic cells, these levels have been shown to promote T-cell recognition [44, 49, 50]. In concurrence with these in vitro studies, multiple reports have indicated that T-cells also mediate acute rejection of PSCs and their derivatives in mice [46, 47, 50–52]. Other studies, however, presented evidences that some PSC derivatives are not targeted by T-cells [45, 53]. Ultimately, these studies indicate that long-term exposure of almost any PSC line or their derivatives to T-cells would ultimately elicit sufficient sensitization for an immune attack. In contrast, in the case of isografts derived from hPSCs (e.g., derived from patient specific hiPSCs), the full MHC match would prevent the development of a T-cell mediated acute immune response. This principle was previously demonstrated by transplantation of SCNT-derived PSC progeny into isogenic animals [54]. In this case, despite mitochondrial antigen mismatches (mitochondria are primarily derived from the ova cytoplasm [55]), T cell response was not observed [56].

Although the primary focus of allorejection studies has been the direct cytotoxic response mediated via CD8 T-cells, recent studies have highlighted the involvement of CD4 helper T-cell subsets in graft rejection and survival. It has been shown that hESC transplants survive better in CD4 null compared with CD8 null mice, yet ultimately both strains rejected the human xenografts [52]. Lui et al. also showed that ablation of CD4 T-cells via systemic anti-CD25 antibody treatment permits survival of mouse ESC grafts in immunocompetent CB/K mice [57] and that inhibition of CD4 T-cells severely dampened the CD8 T-cell activity [58].

These data highlight that host T-cells would likely become central mediators for rejection of differentiated hPSCs, either directly through activation of cytotoxic CD8 T-cells or though indirect exposure of transplanted alloantigens to CD4 T-cells.

The fact that MHC and mHC alleles would match in the hPSC isograft setting would cancel many of the immunologic barriers imposed by allogenic transplantation. However, as discussed above, the expression of aberrant antigens even by isogenic cells is likely to promote isograft rejection by the host's T-cells. In addition, as outlined below, retention of embryonic antigen expression by isografts derived from hPSCs may also promote T-cell mediated rejection.

1.3 Retention of Embryonic Antigens May Lead to T-Cell Mediated Rejection of Isogenic hPSC Derivatives

T-cell variability is driven by random rearrangements of the V(D)J region of the *TCR* gene. A diverse array of T-cells is generated in this fashion, of which some specifically recognize self-antigens. These auto-reactive clones are normally depleted thorough apoptosis in the thymus [59]. To allow tolerance towards somatic antigens expressed outside the thymus, medullary epithelial and dendritic thymic cells express the *AIRE* gene which induces transcription of somatic genes [60]. During human development, the fetal thymus becomes capable of rudimentary support of T-cell selection by approximately 7 weeks gestation [61] and produces the first mature T-cells during week 8 [62]. Since thymic development occurs well over a month after the last pluripotent and early germ layer progenitors have differentiated, T-cells reactive to early embryonic antigens may exist in adults [61]. Therefore, embryonic proteins and glycans expressed by hPSC progeny may elicit immune responses, unless they are ectopically expressed in the thymus by *AIRE*. Partial list includes TRA-1-81, TRA-1-60, SSEA-3, SSEA-4, and SSEA-5 glycans [63–66]. These glycans exist as post-translational modifications on hPSC proteins and lipids, producing isoforms specific to early embryonic development [63, 67, 68]. Since immune responses are known to take place against primitive oncofetal antigens expressed by tumors [69–72], it is probable to assume that the thymus does not support negative selection of all T-cells that are specific for embryonic antigens. For example, developmental pluripotency associated 2 protein (DPPA2), which is normally restricted to the placenta and testis is expressed by a subset of ovarian cancers, and is known to elicit immune responses [72]. This notion is further supported by studies utilizing ESCs as part of a vaccination protocol to promote immune responses against tumors, including those of the colon [69] and lung [73].

The known T-cell response against embryonic antigens raise a concern that even isogenic hPSC derivatives would be targeted if embryonic antigens were not entirely removed. Since it is unlikely that the non-physiological differentiation environments of hPSCs in vitro would precisely recapitulate normal development, some of the embryonic antigens may persist on differentiated progeny. In such a scenario, therapeutic products would induce T-cells responses. The likelihood that this pathway will affect transplantation outcomes is currently difficult to predict because the extent to which embryonic antigens are expressed by hPSC progeny is not known. Still, the detection of immune responses against embryonic antigens [72, 73] and the discovery of aberrant DNA methylation patterns in hiPSCs [15] indicate the high probability of immune response towards residual embryonic antigens. We therefore recommend conducting thorough characterization of retained embryonic antigens on differentiated cells intended for transplantation followed by functional evaluation of their immunogenicity.

1.4 Natural Killer Cell-Mediated Lysis of hPSC Derivatives

As implicated by their name, NK cells are poised to lyse cells, particularly, those exhibiting aberrantly low MHC-I levels, a phenomenon often resulting from viral infections [74, 75]. NK cells monitor MHC-I molecules through the NKG2 and KIR receptors. NK cells then integrate inputs from additional activating and inhibitory stimuli to determine whether the cytotoxic threshold has been reached [75, 76]. This mode of immune surveillance has led to the development of the missing-self hypothesis [77].

Similarly to fetal cells, hPSCs and their early derivatives maintain low levels of HLA molecules relative to somatic cells [2, 3, 42]. The reduced MHC-I presentation, which is arguably an important mechanism that prevents recognition of fetal cells by T and B cells [45], may promote rejection by NK cells [54, 78–81]. During pregnancy, low MHC-I expression by fetal cells does not lead to rejection as a specialized subset of $CD56^{+}$ NK cells interact with the trophoblasts in the endometrium, promoting quiescence of maternal NK cells [82]. In transplantation settings, however, if the engrafted hPSC derivatives will present low MHC-I levels, there exists the probability that they will be probed and lysed by circulating NK cells. Derivatives of parthenogenetic hESCs, in particular, may be at high risk of NK cell-rejection, as lines that are derived from unfertilized metaphase II (MII) oocytes harbor only one set of chromosome homologs [5]. Since MII lines are capable of expressing only one allele of each HLA-gene, such cells are more likely to promote NK cell response due to their inherently reduced HLA expression [83].

Similarly to T-cells, certain NK cells undergo selection or "education". Although this process, termed licensing, is not as well understood as T-cell education, multiple studies have indicated that NK cells initially express an array of inhibiting and activating receptors and their interactions with self-MHC molecules on somatic cells determine whether they mature to become functionally competent [74, 75] (for a comprehensive review refer to Orr et al. [76]). As with all lymphocytes, NK cell "education" can begin only after the development of the immune system, which occurs at a relatively advanced gestational stage. It is likely that the early developmental age of hPSC derivatives may not allow for their participation in NK cell education.

Our initial analysis has demonstrated that hESC derivatives express low levels of inhibitory classical HLA-I and nonclassical

HLA-Ib molecules (HLA-E), and that the levels of these molecules increase during differentiation. In addition, we observed only a basal response of activated NK cells towards undifferentiated and differentiated hESCs [2]. NK cells have been reported to be inhibited by hESC derived mesenchymal cells, at least partially by expressing HLA-G, a protein which plays a pivotal role in promoting maternal NK cell tolerance towards the placenta [53]. Since HLA-G is primarily expressed by trophoblasts [84], it is likely that these cells represent placental mesenchyme. Conversely, other reports have indicated that ESC derivatives are targeted by NK cells [54, 78, 79]. For example, Preynat-Seauve et al. [79] reported that neural cells derived from hESCs are lysed by both T and NK cells. Such discrepancies in NK cell responses towards hPSC derivates could emanate from differences in NK cell subtypes utilized for in vitro lysis assays and their degree of activation. Differences in NK cell properties between mouse and human could also account for some of these disparities, especially since some of the studies have examined the xenogeneic response of mouse NK cell towards hPSC derivatives. Given the complexity of NK cell-target interactions and the existence of stimulatory or inhibitory signals it is not surprising that uncertainty exist as to whether low MHC expression on hPSC-derivatives leads to an increase [46, 50, 81] or decrease [45, 85, 86] in NK cell activity. In addition, each differentiated progeny type may profoundly differ in its capacity to promote NK response.

Take together, the uncertainty regarding the interaction of NK cells and hPSC derivatives leads us to stress that clinical advancements require extensive characterization of NK cell responses towards individual cellular products. Due to the inherent limitations of in vitro assays, development of more clinically relevant in vivo assays should be pursued. In addition, since patient hiPSC derivatives are expected to become a major source for isogenic transplants, their susceptibility to NK cell lysis should be rigorously analyzed. It is also worth noting that it is currently unknown whether therapeutic populations derived from hPSCs are able to reach and maintain expression of MHC-I molecules at adult-like levels, including the various different HLA-I subtypes. We therefore conclude that clinical applications of hPSC derivatives require fundamental analyses of their interactions with NK cells, especially in the context of long-term engraftment.

1.5 Additional Mechanisms Modulating Immune Responses Toward Differentiated hPSCs

Recent studies have indicated that PSCs are capable of modulating immune responses [16, 87] in similar ways to embryos in vivo [82]. Yachimovich-Cohen et al. [88] have found that both mouse and human ESCs secrete Arginase I, which inhibits lymphocyte proliferation and TCR expression. This immunosuppressive activity has been shown to be important for pregnancy [89], particularly for protecting fetal-derived trophoblasts from maternal T-cells

[90, 91]. Other studies have also described the ability of ESCs to inhibit immune cells via secretion of TGF-β [92] and by activating the hemoxygenase I enzyme, which produces anti-inflammatory molecules including biliverdin and carbon monoxide [93].

Additional studies proposed that ESC directly suppress immune cells through Fas ligand (FasL, CD95L) presentation [86, 94–96], a molecular which acts by inducing T-cells apoptosis through interaction with the Fas receptor (CD95). The functions of FasL have been extensively studied in the context of immune-privileged sites such as the placenta, where it plays an important role in inhibiting maternal immune reactions [87, 97]. Although the expression of FasL by PSCs is in line with their early embryonic origin, contradictory evidence has indicated that functional FasL is not present in mouse [98] or human [44] ESCs.

It is important to note that the pathways by which undifferentiated hPSCs modulate immune responses are probably not relevant for clinical translations as transplantation of undifferentiated cells is undesirable since they produce teratomas [66, 99]. Hence, FasL or Arginase I could provide immune protection only if they are expressed by the differentiated therapeutic progenies, a possibility which has not been extensively tested thus far. Therefore, we propose to focus future studies on the potential immunomodulating effects of differentiated cells.

1.6 Mild Immunosuppression Is Expected to Improve Engraftment of Isogenic hPSC Derivatives

As hPSCs and their derivatives may activate immune responses in allogeneic and even isogenic hosts, the development of immunosuppression treatments to mute these responses is also under investigation [52, 79, 100]. For example, Swinjinberg et al. [52] have shown that dual treatment with the potent clinical drugs tacrolimus and sirolimus enable survival of mouse ESCs in immunocompetent mice while monotherapy with either drug was ineffective. Notably, a caveat of such aggressive immunosuppression may be the inhibition of graft maturation and function, as highlighted by Preynat-Seauve et al. [79], who showed that cyclosporine and dexamethasone inhibited neuroprogenitor cell differentiation from hESCs. Since T and NK cell response towards tissues generated from isogenic PSC sources is expected to be mild, we propose that gentler conditioning regimes, such as those provided by antibody perturbations should be developed to prevent isogenic tissue rejection.

Several studies have thus far reported encouraging results for the utility of monoclonal antibodies (mAbs) targeting immune cell receptors to inhibit T [51, 58, 100] and NK [78, 79] cell cytotoxicity. Robertson et al. [51] induced tolerance towards undifferentiated mouse ESCs utilizing non-depleting mAbs against the T-cell co-receptors CD4 and CD8. Similarly, Lui et al. [57] showed that inhibition of CD4 T-cells alone through an anti-CD25 mAb is sufficient to induce tolerance towards undifferentiated mouse ESCs. Grinnemo et al. [58] and Pearl et al. [100] went on to show that costimulation blockade of T and dendritic cells utilizing anti-CD40L, CTLA4, and LFA-1 mAbs enabled teratoma formation from hESCs in immunocompetent mice. In addition, mAbs against the NKG2D receptor were shown to inhibit NK-cell mediated PSC lysis in vitro [78, 79, 81].

An important caveat of the majority of the mAb conditioning studies performed thus far is that they utilized teratoma growth as their primary assay. Teratomas are poor surrogates to assess the engraftment potential of functional and clinically applicable progenitors. Notably, numerous cellular properties, including the molecules involved in immune recognition, are altered during differentiation. It is therefore difficult to predict the survival of differentiated progeny while assaying engraftment of undifferentiated cells. Hence, we highlight that it is essential that continual immune modulation research focus more on enhancing the engraftment of functional hPSC derivatives.

As an alternative for immunosuppression, immunoprivileged sites, including the anterior chamber of the eye, the brain, and the testes [101] could serve as "safe havens" for protecting hPSC-derivatives from rejection. Immune-privileged sites are characterized by a number of mechanisms that disable or suppress immune effector cells [101]. The anterior chamber of the eye is lined by secretory cells that produce various cytokines that suppress helper T-cell activity and promote FAS expression, which in turn, promotes T-cell apoptosis [101, 102]. The brain presents a different set of mechanisms to dampen immune cells, including the endothelial blood–brain barrier, which inhibits entry of immune cells including monocytes and T-cells [103]. Since, many routes of hPSC rejection involve T-cells, which are largely inhibited in immunoprivileged sites, these sites are expected to enhance survival. Still, the full extent to which immunoprivileged sites offer protection for hPSC-progeny remains to be functionally tested.

1.7 Conclusions

Since the derivation of hESCs by Thompson et al. [104], the field of PSC biology and their potential clinical utility has been in a constant state of advancement and change. Although the antigenicity and immunogenicity of hPSCs were initially extensively studied, this research slowed following the demonstration that genetically matched hiPSC lines could be prepared from somatic cells. However, this view was based on the assumption that MHC matching will by itself suffice to prevent immune response towards hiPSC derivatives.

In this review, we highlight the possibility that isogenic hPSC-derived transplants should not be considered non-immunogenic although they are genetically identical to an original tissue donor. Unlike conventional matched organ transplants, differentiated hPSCs are propagated and matured in vitro in an artificial environment that promotes non-physiological proliferation and accelerated specification towards committed cells. The propagation of hPSCs in the presence of animal products may lead to presentation of xenogenic epitopes, and the artificial nutrient environment in vitro may promote aberrations in cell surface antigen presentation. In addition, extensive propagation in vitro may select for genetically abnormal hPSC clones expressing deviant surface molecules. Finally there is an incompletely understood yet real chance that incompletely reprogrammed iPSCs may results in increased immunogenicity. These factors alone may significantly contribute to elevated immunogenicity of differentiated hPSCs. Compared with normal development, differentiation in vitro is substantially faster and likely does not allow for complete removal of embryonic antigens and induction of adult epitopes including optimization of the MHC-I levels. Such aberrant immunological signatures could activate cytotoxic T and NK cells which undergo a complex and developmentally regulated education process from which embryonic antigens are likely excluded. Therefore, retention of embryonic antigens on PSC-derivatives may lead to T and/or NK cell response. In regards to expression of immunosuppressive molecules by hPSCs it is currently unclear whether these factors persist to later stages of differentiation, and therefore, their relevance to transplantation tolerance is uncertain. In summary, these data indicate that differentiated isogenic hPSCs could potentially promote rejection processes unless quality controls and immunosuppression measures are taken.

Nevertheless, we predict that the immune reactions against isogenic hPSC derived transplants will be subtler compared to allogenic responses directed at MHC and mHC mismatches. Therefore, we anticipate that immune reactions against isogenic hPSC-derived transplants could be dampened successfully via targeted therapy. Promising therapies include specific mAbs against receptors found on subsets of immune effector cells, rather than global immunosuppression employed during the course of allograft transplantations. However, we stress that arriving upon immunologic solutions requires a deeper understanding of the mechanisms underlying rejection and tolerance in addition to a better understanding of hPSC biology. The subject of immunogenicity of isogenic hPSC derived graft therefore deserves more attention with special emphasis on rigorous in vivo analysis before we can confidently and successfully utilize these cells for regenerative purposes.

References

1. Nankivell BJ, Alexander SI (2010) Rejection of the kidney allograft. N Engl J Med 363:1451–1462
2. Drukker M, Katz G, Urbach A, Schuldiner M, Markel G et al (2002) Characterization of the expression of MHC proteins in human embryonic stem cells. Proc Natl Acad Sci U S A 99:9864–9869
3. Draper JS, Pigott C, Thomson JA, Andrews PW (2002) Surface antigens of human embryonic stem cells: changes upon differentiation in culture. J Anat 200:249–258

4. Drukker M (2008) Immunological considerations for cell therapy using human embryonic stem cell derivatives.
5. Revazova ES, Turovets NA, Kochetkova OD, Kindarova LB, Kuzmichev LN et al (2007) Patient-specific stem cell lines derived from human parthenogenetic blastocysts. Cloning Stem Cells 9:432–449
6. McCreath KJ, Howcroft J, Campbell KH, Colman A, Schnieke AE et al (2000) Production of gene-targeted sheep by nuclear transfer from cultured somatic cells. Nature 405:1066–1069
7. Takahashi K, Tanabe K, Ohnuki M, Narita M, Ichisaka T et al (2007) Induction of pluripotent stem cells from adult human fibroblasts by defined factors. Cell 131:861–872
8. Drukker M (2004) Immunogenicity of human embryonic stem cells: can we achieve tolerance? Springer Semin Immunopathol 26:201–213
9. Drukker M, Benvenisty N (2004) The immunogenicity of human embryonic stem-derived cells. Trends Biotechnol 22:136–141
10. Martin MJ, Muotri A, Gage F, Varki A (2005) Human embryonic stem cells express an immunogenic nonhuman sialic acid. Nat Med 11:228–232
11. Heiskanen A, Satomaa T, Tiitinen S, Laitinen A, Mannelin S et al (2007) N-glycolylneuraminic acid xenoantigen contamination of human embryonic and mesenchymal stem cells is substantially reversible. Stem Cells 25:197–202
12. Chung TL, Turner JP, Thaker N, Kolle G, Cooper-White JJ et al (2010) Ascorbate promotes epigenetic activation of CD30 in human embryonic stem cells. Stem Cells 28(10):1782–1793
13. Draper JS, Smith K, Gokhale P, Moore HD, Maltby E et al (2004) Recurrent gain of chromosomes 17q and 12 in cultured human embryonic stem cells. Nat Biotechnol 22:53–54
14. Spits C, Mateizel I, Geens M, Mertzanidou A, Staessen C et al (2008) Recurrent chromosomal abnormalities in human embryonic stem cells. Nat Biotechnol 26:1361–1363
15. Kim K, Doi A, Wen B, Ng K, Zhao R et al (2010) Epigenetic memory in induced pluripotent stem cells. Nature 467:285–290
16. Fernandez N, Cooper J, Sprinks M, AbdElrahman M, Fiszer D et al (1999) A critical review of the role of the major histocompatibility complex in fertilization, preimplantation development and feto-maternal interactions. Hum Reprod Update 5: 234–248
17. Lei T, Jacob S, Ajil-Zaraa I, Dubuisson JB, Irion O et al (2007) Xeno-free derivation and culture of human embryonic stem cells: current status, problems and challenges. Cell Res 17:682–688
18. Tangvoranuntakul P, Gagneux P, Diaz S, Bardor M, Varki N et al (2003) Human uptake and incorporation of an immunogenic nonhuman dietary sialic acid. Proc Natl Acad Sci U S A 100:12045–12050
19. Chou HH, Takematsu H, Diaz S, Iber J, Nickerson E et al (1998) A mutation in human CMP-sialic acid hydroxylase occurred after the Homo-Pan divergence. Proc Natl Acad Sci U S A 95:11751–11756
20. Cerdan C, Bendall SC, Wang L, Stewart M, Werbowetski T et al (2006) Complement targeting of nonhuman sialic acid does not mediate cell death of human embryonic stem cells. Nat Med 12:1113–1114, author reply 1115
21. Ludwig TE, Levenstein ME, Jones JM, Berggren WT, Mitchen ER et al (2006) Derivation of human embryonic stem cells in defined conditions. Nat Biotechnol 24: 185–187
22. Mallon BS, Park KY, Chen KG, Hamilton RS, McKay RD (2006) Toward xeno-free culture of human embryonic stem cells. Int J Biochem Cell Biol 38:1063–1075
23. Xu C, Inokuma MS, Denham J, Golds K, Kundu P et al (2001) Feeder-free growth of undifferentiated human embryonic stem cells. Nat Biotechnol 19:971–974
24. Prowse AB, Doran MR, Cooper-White JJ, Chong F, Munro TP et al (2010) Long term culture of human embryonic stem cells on recombinant vitronectin in ascorbate free media. Biomaterials 31:8281–8288
25. Rajala K, Hakala H, Panula S, Aivio S, Pihlajamaki H et al (2007) Testing of nine different xeno-free culture media for human embryonic stem cell cultures. Hum Reprod 22:1231–1238
26. Newman AM, Cooper JB (2010) Lab-specific gene expression signatures in pluripotent stem cells. Cell Stem Cell 7:258–262
27. Lefort N, Feyeux M, Bas C, Feraud O, Bennaceur-Griscelli A et al (2008) Human embryonic stem cells reveal recurrent genomic instability at 20q11.21. Nat Biotechnol 26:1364–1366
28. Werbowetski-Ogilvie TE, Bosse M, Stewart M, Schnerch A, Ramos-Mejia V et al (2009) Characterization of human embryonic stem cells with features of neoplastic progression. Nat Biotechnol 27:91–97
29. Herszfeld D, Wolvetang E, Langton-Bunker E, Chung TL, Filipczyk AA et al (2006) CD30 is a survival factor and a biomarker for transformed human pluripotent stem cells. Nat Biotechnol 24:351–357

30. Al-Shamkhani A (2004) The role of CD30 in the pathogenesis of haematopoietic malignancies. Curr Opin Pharmacol 4:355–359
31. Marchetto MC, Yeo GW, Kainohana O, Marsala M, Gage FH et al (2009) Transcriptional signature and memory retention of human-induced pluripotent stem cells. PLoS One 4:e7076
32. Zhao T, Zhang ZN, Rong Z, Xu Y (2011) Immunogenicity of induced pluripotent stem cells. Nature 474:212–215
33. Stelzer Y, Yanuka O, Benvenisty N (2011) Global analysis of parental imprinting in human parthenogenetic induced pluripotent stem cells. Nat Struct Mol Biol 18:735–741
34. Guenther MG, Frampton GM, Soldner F, Hockemeyer D, Mitalipova M et al (2010) Chromatin structure and gene expression programs of human embryonic and induced pluripotent stem cells. Cell Stem Cell 7:249–257
35. Hochedlinger K, Jaenisch R (2003) Nuclear transplantation, embryonic stem cells, and the potential for cell therapy. N Engl J Med 349:275–286
36. Adewumi O, Aflatoonian B, Ahrlund-Richter L, Amit M, Andrews PW et al (2007) Characterization of human embryonic stem cell lines by the International Stem Cell Initiative. Nat Biotechnol 25:803–816
37. Moore JC, Sadowy S, Alikani M, Toro-Ramos AJ, Swerdel MR et al (2010) A high-resolution molecular-based panel of assays for identification and characterization of human embryonic stem cell lines. Stem Cell Res 4:92–106
38. Akopian V, Andrews PW, Beil S, Benvenisty N, Brehm J et al (2010) Comparison of defined culture systems for feeder cell free propagation of human embryonic stem cells. In Vitro Cell Dev Biol Anim 46:247–258
39. Zeng F, Zhou Q, Tannenbaum S (2009) Consensus guidance for banking and supply of human embryonic stem cell lines for research purposes. Stem Cell Rev 5:301–314
40. Sprinks MT, Sellens MH, Dealtry GB, Fernandez N (1993) Preimplantation mouse embryos express Mhc class I genes before the first cleavage division. Immunogenetics 38: 35–40
41. Cooper JC, Fernandez N, Joly E, Dealtry GB (1998) Regulation of major histocompatibility complex and TAP gene products in preimplantation mouse stage embryos. Am J Reprod Immunol 40:165–171
42. Suarez-Alvarez B, Rodriguez RM, Calvanese V, Blanco-Gelaz MA, Suhr ST et al (2010) Epigenetic mechanisms regulate MHC and antigen processing molecules in human embryonic and induced pluripotent stem cells. PLoS One 5:e10192
43. Jurisicova A, Casper RF, MacLusky NJ, Mills GB, Librach CL (1996) HLA-G expression during preimplantation human embryo development. Proc Natl Acad Sci U S A 93: 161–165
44. Drukker M, Katchman H, Katz G, Even-Tov Friedman S, Shezen E et al (2006) Human embryonic stem cells and their differentiated derivatives are less susceptible to immune rejection than adult cells. Stem Cells 24:221–229
45. Mammolenti M, Gajavelli S, Tsoulfas P, Levy R (2004) Absence of major histocompatibility complex class I on neural stem cells does not permit natural killer cell killing and prevents recognition by alloreactive cytotoxic T lymphocytes in vitro. Stem Cells 22:1101–1110
46. Wu DC, Boyd AS, Wood KJ (2008) Embryonic stem cells and their differentiated derivatives have a fragile immune privilege but still represent novel targets of immune attack. Stem Cells 26:1939–1950
47. Boyd AS, Wood KJ (2009) Variation in MHC expression between undifferentiated mouse ES cells and ES cell-derived insulin-producing cell clusters. Transplantation 87:1300–1304
48. Lampton PW, Crooker RJ, Newmark JA, Warner CM (2008) Expression of major histocompatibility complex class I proteins and their antigen processing chaperones in mouse embryonic stem cells from fertilized and parthenogenetic embryos. Tissue Antigens 72:448–457
49. Utermohlen O, Baschuk N, Abdullah Z, Engelmann A, Siebolts U et al (2009) Immunologic hurdles of therapeutic stem cell transplantation. Biol Chem 390:977–983
50. Dressel R, Guan K, Nolte J, Elsner L, Monecke S et al (2009) Multipotent adult germ-line stem cells, like other pluripotent stem cells, can be killed by cytotoxic T lymphocytes despite low expression of major histocompatibility complex class I molecules. Biol Direct 4:31
51. Robertson NJ, Brook FA, Gardner RL, Cobbold SP, Waldmann H et al (2007) Embryonic stem cell-derived tissues are immunogenic but their inherent immune privilege promotes the induction of tolerance. Proc Natl Acad Sci U S A 104:20920–20925
52. Swijnenburg RJ, Schrepfer S, Govaert JA, Cao F, Ransohoff K et al (2008) Immunosuppressive therapy mitigates immunological rejection of human embryonic stem cell xenografts. Proc Natl Acad Sci U S A 105:12991–12996
53. Yen BL, Chang CJ, Liu KJ, Chen YC, Hu HI et al (2009) Brief report–human embryonic stem cell-derived mesenchymal progenitors possess strong immunosuppressive effects toward natural killer cells as well as T lymphocytes. Stem Cells 27:451–456

54. Rideout WM III, Hochedlinger K, Kyba M, Daley GQ, Jaenisch R (2002) Correction of a genetic defect by nuclear transplantation and combined cell and gene therapy. Cell 109:17–27
55. Evans MJ, Gurer C, Loike JD, Wilmut I, Schnieke AE et al (1999) Mitochondrial DNA genotypes in nuclear transfer-derived cloned sheep. Nat Genet 23:90–93
56. Lanza RP, Chung HY, Yoo JJ, Wettstein PJ, Blackwell C et al (2002) Generation of histocompatible tissues using nuclear transplantation. Nat Biotechnol 20:689–696
57. Lui KO, Boyd AS, Cobbold SP, Waldmann H, Fairchild PJ (2010) A role for regulatory T cells in acceptance of embryonic stem cell-derived tissues transplanted across an MHC barrier. Stem Cells 28(10):1905–1914
58. Grinnemo KH, Genead R, Kumagai-Braesch M, Andersson A, Danielsson C et al (2008) Costimulation blockade induces tolerance to HESC transplanted to the testis and induces regulatory T-cells to HESC transplanted into the heart. Stem Cells 26:1850–1857
59. Nitta T, Murata S, Ueno T, Tanaka K, Takahama Y (2008) Thymic microenvironments for T-cell repertoire formation. Adv Immunol 99:59–94
60. Gardner JM, Fletcher AL, Anderson MS, Turley SJ (2009) AIRE in the thymus and beyond. Curr Opin Immunol 21:582–589
61. Res P, Spits H (1999) Developmental stages in the human thymus. Semin Immunol 11:39–46
62. Haynes BF, Heinly CS (1995) Early human T cell development: analysis of the human thymus at the time of initial entry of hematopoietic stem cells into the fetal thymic microenvironment. J Exp Med 181:1445–1458
63. Nagano K, Yoshida Y, Isobe T (2008) Cell surface biomarkers of embryonic stem cells. Proteomics 8:4025–4035
64. Kannagi R, Cochran NA, Ishigami F, Hakomori S, Andrews PW et al (1983) Stage-specific embryonic antigens (SSEA-3 and -4) are epitopes of a unique globo-series ganglioside isolated from human teratocarcinoma cells. EMBO J 2:2355–2361
65. Shevinsky LH, Knowles BB, Damjanov I, Solter D (1982) Monoclonal antibody to murine embryos defines a stage-specific embryonic antigen expressed on mouse embryos and human teratocarcinoma cells. Cell 30:697–705
66. Tang C, Lee AS, Volkmer JP, Sahoo D, Nag D et al (2011) An antibody against SSEA-5 glycan on human pluripotent stem cells enables removal of teratoma-forming cells. Nat Biotechnol 29(9):829–834
67. Schopperle WM, DeWolf WC (2007) The TRA-1-60 and TRA-1-81 human pluripotent stem cell markers are expressed on podocalyxin in embryonal carcinoma. Stem Cells 25:723–730
68. Brimble SN, Sherrer ES, Uhl EW, Wang E, Kelly S et al (2007) The cell surface glycosphingolipids SSEA-3 and SSEA-4 are not essential for human ESC pluripotency. Stem Cells 25:54–62
69. Li Y, Zeng H, Xu RH, Liu B, Li Z (2009) Vaccination with human pluripotent stem cells generates a broad spectrum of immunological and clinical responses against colon cancer. Stem Cells 27:3103–3111
70. Siegel S, Wagner A, Kabelitz D, Marget M, Coggin J Jr et al (2003) Induction of cytotoxic T-cell responses against the oncofetal antigen-immature laminin receptor for the treatment of hematologic malignancies. Blood 102:4416–4423
71. Zelle-Rieser C, Barsoum AL, Sallusto F, Ramoner R, Rohrer JW et al (2001) Expression and immunogenicity of oncofetal antigen-immature laminin receptor in human renal cell carcinoma. J Urol 165:1705–1709
72. Tchabo NE, Mhawech-Fauceglia P, Caballero OL, Villella J, Beck AF et al (2009) Expression and serum immunoreactivity of developmentally restricted differentiation antigens in epithelial ovarian cancer. Cancer Immun 9:6
73. Dong W, Du J, Shen H, Gao D, Li Z et al (2010) Administration of embryonic stem cells generates effective antitumor immunity in mice with minor and heavy tumor load. Cancer Immunol Immunother 59:1697–1705
74. Jonsson AH, Yokoyama WM (2009) Natural killer cell tolerance licensing and other mechanisms. Adv Immunol 101:27–79
75. Yokoyama WM, Kim S (2006) How do natural killer cells find self to achieve tolerance? Immunity 24:249–257
76. Orr MT, Lanier LL (2010) Natural killer cell education and tolerance. Cell 142:847–856
77. Borrego F (2006) The first molecular basis of the "missing self" hypothesis. J Immunol 177:5759–5760
78. Frenzel LP, Abdullah Z, Kriegeskorte AK, Dieterich R, Lange N et al (2009) Role of natural-killer group 2 member D ligands and intercellular adhesion molecule 1 in natural killer cell-mediated lysis of murine embryonic stem cells and embryonic stem cell-derived cardiomyocytes. Stem Cells 27:307–316
79. Preynat-Seauve O, de Rham C, Tirefort D, Ferrari-Lacraz S, Krause KH et al (2009) Neural progenitors derived from human embryonic stem cells are targeted by allogeneic T and natural killer cells. J Cell Mol Med 13:3556–3569

80. Dressel R, Schindehutte J, Kuhlmann T, Elsner L, Novota P et al (2008) The tumorigenicity of mouse embryonic stem cells and in vitro differentiated neuronal cells is controlled by the recipients' immune response. PLoS One 3:e2622
81. Dressel R, Nolte J, Elsner L, Novota P, Guan K et al (2010) Pluripotent stem cells are highly susceptible targets for syngeneic, allogeneic, and xenogeneic natural killer cells. FASEB J 24(7):2164–2177
82. Trowsdale J, Betz AG (2006) Mother's little helpers: mechanisms of maternal–fetal tolerance. Nat Immunol 7:241–246
83. Revazova ES, Turovets NA, Kochetkova OD, Agapova LS, Sebastian JL et al (2008) HLA homozygous stem cell lines derived from human parthenogenetic blastocysts. Cloning Stem Cells 10:11–24
84. Kovats S, Main EK, Librach C, Stubblebine M, Fisher SJ et al (1990) A class I antigen, HLA-G, expressed in human trophoblasts. Science 248:220–223
85. Magliocca JF, Held IK, Odorico JS (2006) Undifferentiated murine embryonic stem cells cannot induce portal tolerance but may possess immune privilege secondary to reduced major histocompatibility complex antigen expression. Stem Cells Dev 15:707–717
86. Li L, Baroja ML, Majumdar A, Chadwick K, Rouleau A et al (2004) Human embryonic stem cells possess immune-privileged properties. Stem Cells 22:448–456
87. Guller S, LaChapelle L (1999) The role of placental Fas ligand in maintaining immune privilege at maternal–fetal interfaces. Semin Reprod Endocrinol 17:39–44
88. Yachimovich-Cohen N, Even-Ram S, Shufaro Y, Rachmilewitz J, Reubinoff B (2010) Human embryonic stem cells suppress T cell responses via arginase I-dependent mechanism. J Immunol 184:1300–1308
89. Bronte V, Zanovello P (2005) Regulation of immune responses by L-arginine metabolism. Nat Rev Immunol 5:641–654
90. Ishikawa T, Harada T, Koi H, Kubota T, Azuma H et al (2007) Identification of arginase in human placental villi. Placenta 28:133–138
91. Kropf P, Baud D, Marshall SE, Munder M, Mosley A et al (2007) Arginase activity mediates reversible T cell hyporesponsiveness in human pregnancy. Eur J Immunol 37:935–945
92. Koch CA, Geraldes P, Platt JL (2008) Immunosuppression by embryonic stem cells. Stem Cells 26:89–98
93. Trigona WL, Porter CM, Horvath-Arcidiacono JA, Majumdar AS, Bloom ET (2007) Could heme-oxygenase-1 have a role in modulating the recipient immune response to embryonic stem cells? Antioxid Redox Signal 9:751–756
94. Fandrich F, Lin X, Chai GX, Schulze M, Ganten D et al (2002) Preimplantation-stage stem cells induce long-term allogeneic graft acceptance without supplementary host conditioning. Nat Med 8:171–178
95. Fabricius D, Bonde S, Zavazava N (2005) Induction of stable mixed chimerism by embryonic stem cells requires functional Fas/FasL engagement. Transplantation 79: 1040–1044
96. Bonde S, Zavazava N (2006) Immunogenicity and engraftment of mouse embryonic stem cells in allogeneic recipients. Stem Cells 24:2192–2201
97. Hunt JS, Vassmer D, Ferguson TA, Miller L (1997) Fas ligand is positioned in mouse uterus and placenta to prevent trafficking of activated leukocytes between the mother and the conceptus. J Immunol 158: 4122–4128
98. Brunlid G, Pruszak J, Holmes B, Isacson O, Sonntag KC (2007) Immature and neurally differentiated mouse embryonic stem cells do not express a functional Fas/Fas ligand system. Stem Cells 25:2551–2558
99. Lee AS, Tang C, Cao F, Xie X, van der Bogt K et al (2009) Effects of cell number on teratoma formation by human embryonic stem cells. Cell Cycle 8:2608–2612
100. Pearl JI, Lee AS, Leveson-Gower DB, Sun N, Ghosh Z et al (2011) Short-term immunosuppression promotes engraftment of embryonic and induced pluripotent stem cells. Cell Stem Cell 8:309–317
101. Matzinger P, Kamala T (2011) Tissue-based class control: the other side of tolerance. Nat Rev Immunol 11:221–230
102. Ferguson TA, Griffith TS (2006) A vision of cell death: Fas ligand and immune privilege 10 years later. Immunol Rev 213: 228–238
103. Engelhardt B, Coisne C (2011) Fluids and barriers of the CNS establish immune privilege by confining immune surveillance to a two-walled castle moat surrounding the CNS castle. Fluids Barriers CNS 8:4
104. Thomson JA, Itskovitz-Eldor J, Shapiro SS, Waknitz MA, Swiergiel JJ et al (1998) Embryonic stem cell lines derived from human blastocysts. Science 282:1145–1147

Chapter 3

Hematopoietic and Nature Killer Cell Development from Human Pluripotent Stem Cells

Zhenya Ni, David A. Knorr, and Dan S. Kaufman

Abstract

Natural killer (NK) cells are key effectors of the innate immune system, protecting the host from a variety of infections, as well as malignant cells. Recent advances in the field of NK cell biology have led to a better understanding of how NK cells develop. This progress has directly translated to improved outcomes in patients receiving hematopoietic stem cell transplants to treat potentially lethal malignancies. However, key differences between mouse and human NK cell development and biology limits the use of rodents to attain a more in depth understanding of NK cell development. Therefore, a readily accessible and genetically tractable cell source to study human NK cell development is warranted. Our lab has pioneered the development of lymphocytes, specifically NK cells, from human embryonic stem cells (hESCs) and more recently induced pluripotent stem cells (iPSCs). This chapter describes a reliable method to generate NK cells from hESCs and iPSCs using murine stromal cell lines. Additionally, we include an updated approach using a spin-embryoid body (spin-EB) differentiation system that allows for human NK cell development completely defined in vitro conditions.

Key words Embryonic stem cells, Induced pluripotent stem cells, Stromal cell coculture, Embryoid body, Hematopoietic progenitors, Natural killer cells

1 Introduction

hESCs and iPSCs provide important starting cell populations to develop new cell-based therapies to treat both malignant and non-malignant hematological diseases. However, the advancement of these treatments depends on a better understanding of the normal development and physiology of desired cell populations. Our group has previously demonstrated the ability of hESC-derived hematopoietic progenitor cells to produce functional NK cells [1–3]. Studies using hESCs or iPSCs which can be readily genetically engineered [4, 5] provide new model systems to study human lymphocyte development and to produce enhanced cell-based therapies. hESC-derived NK cells could serve as a "universal" source of anti-tumor lymphocytes for novel clinical therapies. With continued advances in the stem cell field, it is likely that

Nicholas Zavazava (ed.), *Embryonic Stem Cell Immunobiology: Methods and Protocols*, Methods in Molecular Biology, vol. 1029, DOI 10.1007/978-1-62703-478-4_3,

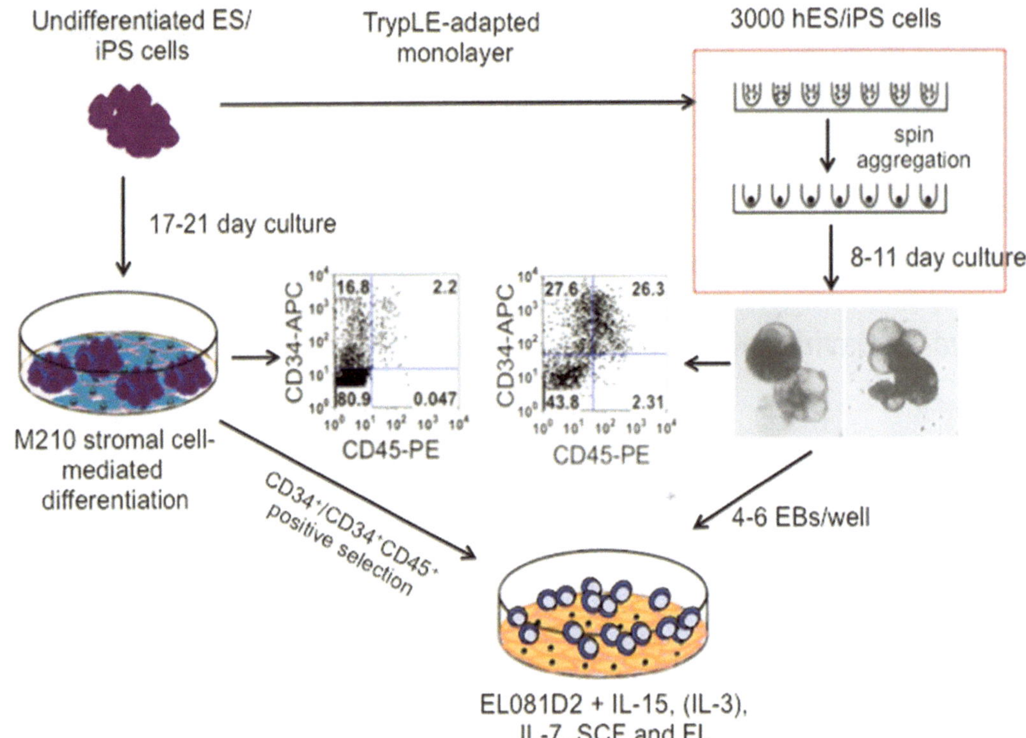

Fig. 1 Schematic diagram of two-step hematopoietic and NK cell differentiation from human embryonic stem cells or induced pluripotent stem cells. Undifferentiated hESC/iPSCs can be induced to differentiate into hematopoietic progenitors either by coculture with M210-B4 stromal cells or by spin embryoid body formation. The hematopoietic progenitors can be characterized based on expression of specific cell surface markers (e.g., CD34, CD45). To generate NK cells, $CD34^{+}CD45^{+}$ progenitors are enriched from M210-B4 coculture and cultured on EL08-1D2 stromal cells in medium supplemented with cytokines supporting NK cell differentiation. Differentiated spin EBs can be directly transferred onto El08-1D2 cells in medium plus same cytokines. After 4–5 weeks, mature and functional NK cells typically develop and can be analyzed for phenotype and function

iPSC-derived NK cells will soon be able to be efficiently derived on a patient-specific basis.

hESC and iPSC-derived NK cells express activating and inhibitory receptors similar to NK cells isolated from adult peripheral blood [3]. The hESC-derived NK cells are also highly efficient at direct cell-mediated cytotoxicity and antibody dependent cell-mediated cytotoxicity, as well as cytokine (IFN-γ) production. Our lab has now developed two separate protocols (stroma or stroma-free) to generate hematopoietic progenitors, both of which are capable of forming functional NK cells. Both systems have their inherent advantages and disadvantages. We traditionally used coculture with murine stroma to support hematopoietic differentiation because the embryoid body (EB) approaches were more variable. However, we have recently adapted use of a “spin EB” protocol [6, 7] to provide as system for more consistent blood cell differentiation without the use of murine stroma (Fig. 1).

2 Materials

2.1 Cell Lines

1. H9 line of hESCs (WiCell, Madison, WI) and an iPSC line generated in our lab by transfection of $CD34^+$ umbilical cord blood cells with four transcription factors Oct4, Sox2, c-Myc, and Klf4. H9 and iPS cells are maintained as undifferentiated cells as previously described [8].
2. M210-B4 stromal cells (American Type Culture Collection, Manassas, VA). Maintained in Dulbecco's modified Eagle's medium (DMEM) High Glucose (Invitrogen Corporation/Gibco; cat. no. 11965-092) supplemented with 15 % defined fetal bovine serum (FBS, Hyclone; cat. no. SH30070.03), 0.1 mM β-mercaptoethanol, 2 mM L-glutamine, 1 % MEM nonessential amino acids solution, and 1 % penicillin-streptomycin (P/S, Invitrogen Corporation/Gibco; cat. no. 15140-122).
3. EL08-1D2 stromal cells (kindly provided by Drs. Rob Oostendorp and Majlinda Lako) [9]. Maintained in 50 % Myelocult M5300 (Stem Cell Tech. cat. no. 05350), 35 % Alpha MEM (Invitrogen cat. no. 12571-063), 15 % FBS (Stem Cell Tech. cat. no. 06500), 1 % Glutamax 1 (100×, Invitrogen cat. no. 35050-061), 0.1 mM β-mercaptoethanol, 10^{-6} M hydrocortisone (Stem Cell Tech. cat. no. 07904), and 1 % P/S.

2.2 Hematopoietic Differentiation of Human ES/iPS Cells on Murine Stromal Cells

1. hESC or iPSC/M210-B4 Differentiation Medium (R-15): RMPI 1640 (Cellgro/Mediatech;cat.no. 10-040-CV) medium containing 15 % fetal bovine serum (FBS) certified (Invitrogen Corporation/Gibco; cat. no. 16000-044), 0.1 mM β-mercaptoethanol (Sigma; cat. no. M75222), 2 mM L-glutamine (Invitrogen Corporation/Gibco; cat. no. 21051-024); 1 % MEM nonessential amino acids solution, and 1 % P/S.
2. D-10 Medium used for wash: DMEM supplemented with 10 % FBS plus 1 % P/S.
3. Collagenase split medium: DMEM/F12 medium containing 1 mg/mL collagenase type IV (Invitrogen Corporation/Gibco; cat. no. 17104-019). Collagenase medium is filter sterilized with a 50 mL, 0.22 μm membrane Steriflip (Millipore; cat. no. SCGP00525).
4. Trypsin-ethylene diamine tetraacetic acid (EDTA)+2 % chicken serum: 0.05 % tryspin 0.53 mM EDTA solution (Cellgro/Mediatech; cat. no. 25-052-CI) with 2 % chicken serum (Sigma; cat. no. C5405).
5. 6-Well tissue culture plates (NUNC Brand Products, Nalgene NUNC cat. no. 152795).
6. Gelatin (Sigma, cat. no. G-1890), 0.1 % solution made in water and sterilized by autoclaving.

2.3 Enrichment of Hematopoietic Progenitors CD34$^+$/CD34$^+$45$^+$ from Stromal Coculture

1. EasySep® human CD34 positive selection kit (StemCell Technologies, cat.no. 18056) and EasySep® PE positive selection kit (StemCell Technologies, cat.no. 18557).
2. EasySep buffer: Dulbecco's Phosphate Buffered Saline (DPBS) without Ca^{2+} and Mg^{2+} (supplemented with 2 % FBS and 1 mM EDTA. Store at 4 °C.
3. 0.05 % Trypsin-EDTA (Cellgro/Mediatech. cat. no. 25-052-CI) + 2 % chick serum.
4. 70 μm Cell strainer filter (Becton Dickinson/Falcon, cat. no. 352350).
5. PE-conjugated anti-human CD45 (BD Pharmingen, cat. no. 555483).

2.4 Hematopoietic Differentiation of hES/iPS Cells from Spin EBs

1. Spin EB differentiation medium (BPEL media): 43 % Iscove's Modified Dulbecco's Medium (IMDM, Thermo, cat. no. SH30228.01), 43 % F-12 Nutrient Mixture w/Glutamax I (Invitrogen, cat. no. 3176503), 0.25 % Deionized bovine serum albumin (BSA, Sigma Aldrich cat. no. A3311), 0.25 % Poly (vinyl alcohol) (Sigma-Aldrich cat. no. P8136), 0.1 μg/mL Linoleic Acid (Sigma-Aldrich cat. no. L1012), 0.1 μg/mL Linolenic Acid (Sigma-Aldrich cat. no. L2376), 1:500 Synthechol 500× solution (Sigma-Aldrich cat. no. S5442), 450 μM α-monothioglycerol (α-MTG) (Sigma-Aldrich cat. no. M6145), 5 % Protein-free hybridoma mix II (Invitrogen cat. no. 12040077), 50 μg/mL Ascorbic acid 2-phosphate (Sigma-Aldrich cat. no. A8960), 2 mM Glutamax I (Invitrogen cat. no. 35050061), 1 % Insulin, Transferrin, Selenium 100× solution (ITS) (Invitrogen cat. no. 41400-045), 1 % Pen/Strep plus 40 μg/mL recombinant human stem cell factor (SCF) (PeproTech cat. no. 300-07), 20 μg/mL BMP4 (R&D systems, cat. no. 314-BP), and 20 μg/mL VEGF (R&D systems, cat. no. 293-VE).
2. TrypLE Select (Gibco/Invitrogen, cat. no. 12563-011).
3. 96-Well round bottom plates (NUNC cat. no. 262162 with lids, cat. no. 264122).

2.5 Nature Killer Cell Differentiation from Enriched CD34$^+$/CD34$^+$45$^+$ Hematopoietic Progenitors or Unsorted Spin EBs

1. NK cell development medium: 56.6 % DMEM-high glucose, 28.3 % HAMS/F12 (Invitrogen, cat. no. 11765-064), 15 % heat-inactivated human AB serum (Valley Biomedical, cat. no. HP1022 HI), 2 mM L-glutamine, 25 μM β-mercaptoethanol, 5 ng/mL sodium selenite (Sigma-Aldrich, cat. no. S5261), 50 μM ethanolamine (MP Biomedicals, cat. no. 194658), 20 mg/L ascorbic acid (Sigma-Aldrich, cat. no. A-5960), 1 % P/S, 5 ng/mL IL-3 (PeproTech, cat. no. 200-03), 20 ng/mL

SCF, 20 ng/mL IL-7 (PeproTech, cat. no.), 10 ng/mL IL-15 (PeproTech, cat. no. 200-15), 10 ng/mL Flt3 ligand (Flt3L) (PeproTech, cat. no.300-19). Store at 4 °C in the dark.

2. 24-Well tissue culture plates (NUNCTM Brand Products, Nalgene Nunc; cat. no. 142475).

3 Methods

3.1 Generation of Hematopoietic Progenitors from hESCs or iPSCs Using Stromal Cell Coculture

1. Culture of Undifferentiated hESCs/iPSCs.

   The confluency of undifferentiated hESCs/iPSCs should be approximately 90–95 % by the day they are passed onto M210-B4 stromal cells for hematopoietic differentiation.

2. Preparation of M210-B4 Feeder Layer.

   Mouse bone marrow M210-B4 cells are maintained in M210-B4 culture medium. To prepare feeder layers, the M210-B4 cells are treated with 10 μg/mL mitomycin C for 4 h at 37 °C, 5 % CO_2. After mitomycin C treatment cells are washed with DPBS three times and then dissociated with 0.05 % trypsin-EDTA to make a single suspension. Inactivated M210-B4s are plated onto 0.1 % gelatin-coated six-well plates with 2.5 mL/well at 1.0×10^5 cells/mL (*see* **Note 1**).

3. Coculture of hESCs/iPSCs on M210-B4 stromal cells.

   Undifferentiated hESCs/iPSCs are dislodged and disrupted by incubation in 1 mg/mL collagenase type IV for approximately 5 min at 37 °C, followed by scraping colonies off with a 5 mL glass pipette. Cells are washed with D-10 medium twice to remove residual collagenase (*see* **Note 2**).

4. Disrupted hES/iPS colonies are then resuspended in R-15 differentiation medium and transferred to M210-B4 monolayer at 1: 4. Cocultures are incubated at 37 °C, 5 % CO_2 for 17–21 days with medium change every 2–3 days (*see* **Note 3**).

3.2 Enrichment of Hematopoietic Progenitors from M210-B4 Coculture

Optimal time for harvesting hematopoietic progenitors varies somehow depending on the hESC/iPSC line and stromal cells and serum lots. However, 10–15 % $CD34^+$ hESC/iPSC-derived cells are generally observed by flow cytometry after 17 days of coculture with M210-B4s (*see* **Note 3**). A time course experiment in which cells are sampled every 2–3 days is recommended to find the best time for enrichment of hematopoietic progenitors as assayed by flow cytometry and/or colony forming assays in methylcellulose [10]. Further sorting for CD34 and CD45 double positive cells can increase the kinetics and frequency of NK cell differentiation [1].

1. To prepare a single-cell suspension from differentiated hES/iPS cells, aspirate medium from all six-well plates of cultures

and add 1 mg/mL collagenase IV 1.5 mL per well for 5–7 min at 37 °C until stromal layer can be observed to break/peel up. Scrape with a 5 mL glass pipette to disrupt the cells and transfer to a 50 mL conical tube. Add 5 mL of Ca^{2+} and Mg^{2+}-free PBS and mix by pipetting up and down against the bottom of the tube until there is a fine suspension of cells is observed. Centrifuge cell suspension at 400 × *g* for 5 min.

2. Remove the supernatant and add 5 mL of 0.05 % prewarmed trypsin/EDTA + 2 % chick serum solution into the tube. Mix the cell suspension by vigorously pipetting up and down. Incubate tube in 37 °C water bath for 5–10 min until a single cell suspension can be observed. It is recommended to shake the tube at regular intervals during the incubation (*see* **Note 4**).
3. Add 10 mL ES wash medium and mix by pipetting to further disperse cells. Centrifuge at 400 × *g* for 5 min. Resuspend cells in 5 mL ES wash medium. Filter the cell suspension with 70 μm cell strainer filter to remove any remaining clumps of cells. Wash the filter once with ES wash medium.
4. Count live cells by 0.4 % tryphan blue staining using a hemacytometer. In general, $1–2 \times 10^6$ single cells can be obtained from a nearly confluent well.
5. Isolate $CD34^+$/$CD34^+CD45^+$ hES or iPS-derived hematopoietic progenitors by using a EasySep® CD34 positive selection kit followed by CD45-PE staining of enriched $CD34^+$ cells and a second enrichment for CD45-PE positive cells using a PE-positive selection kit according to manufacturer's instructions.
6. Count live cells and distribute to 24-well EL08-1D2 plate at desired concentration for NK cell differentiation.

3.3 Generation of Hematopoietic Progenitors from hES/iPS Cells by Spin-EB Formation

Many studies have investigated hematopoietic differentiation by embryoid body (EB) formation, the advantages of this way are defined culture conditions and higher efficiency of hematopoietic progenitor generation. However, EB system cultures that maintain cells in suspension can be variable and cell yields at typically lower compared with the coculture system. In contrast, the spin-EB approach [6, 11] allows a more consistent generation of hematopoietic progenitors that actively proliferate in culture.

1. TrypLE-adapted undifferentiated hESCs/iPSCs are maintained on low density MEF feeders as described [6, 11]. One or two days before setting up Spin EB differentiation, pass TrypLE-adapted hESCs/iPSCs onto fresh MEFs at 1:1 ratio that will allow them to be 80–90 % confluent on the day of differentiation setup.

2. To prepare for Spin EB plating, pipet 150 μL sterile water into the 36 outer wells of each 96-well plate to minimize loss of well volume to evaporation.
3. Aspirate culture media from hESCs/iPSCs and add 1.0 mL pre-warmed TrypLE Select to each well. Place plates in incubator (37 °C) until ES cells start to come off the plate. Typically takes ~5 min if TrypLE prewarmed; do not leave longer than 5 min.
4. Collect dissociated cells in a conical tube and pipet up and down to break up clumps. Dilute TrypLE with 1 volume BPEL media and at least 1 volume DPBS. Spin cells down at 1,500 rpm, 5 min, 8 °C. Remove supernatant and resuspend the cells in 5 mL BPEL media plus 5 mL DPBS. Spin cells down again.
5. Remove supernatant and resuspend cells in 5–10 mL BPEL media. Pass cells through 70 μm filter into a fresh 50 mL conical in order to remove clumps. Count filtered cells and aliquote cells to be used for plating into a 50 mL conical. Spin cells down and resuspend them with stage I spin EB differentiation medium to 3×10^4 cells/mL.
6. Transfer 100 μL cell aliquots into each of the inner 60 wells of the prepared 96-well plates with 150 μl of water in outer wells. Spin 96-well plates at $480 \times g$, 8 °C for 4 min. Incubate the plates at 37 °C, 5 % CO_2 for 8–11 days till hematopoietic progenitor $CD34^+CD45^+$ cells are generated. Do not disturb the plates during the first 3 days of differentiation while the EBs are forming (*see* **Note 5**). Under optimal conditions, the percentage of $CD34^+$ cells can be approximately 40–60 % and percentage of $CD34^+CD45^+$ cells can be up to 20–40 %. By day 11, the good differentiation should have more hematopoietic cells surrounding the initial EBs. Most cells remaining in the EB are endothelial/mesenchymal progenitor populations.

3.4 Natural Killer Cell Differentiation

1. The EL08-1D2 feeder plates are usually prepared the day before needed. Dissociate cells from T-75 flask with 0.05 % trypsin. Spin down and resuspend cells in 80 % fresh medium +20 % old medium to $1.2–1.25 \times 10^5$ cells/mL. Cell are plated into 0.1 % gelatin coated-24-well plated at 1 mL/well and grow to 85–90 % confluency at 33 °C, 5 % CO_2 by next day. After X-ray irradiated at 3,000 rads, the feeders are ready for use or can be saved up to 3–4 days (*see* **Note 6**).
2. Plate isolated hematopoietic progenitors from M210-B4 cocultures onto EL08-1D2 feeders at 10,000–50,000 cells per well in 1 mL NK cell differentiation medium depending on the purity of $CD34^+CD45^+$ population. For NK cell differentiation

from spin EBs, whole spin EBs (without sorting) are directly transferred 24-well EL08-1D2 plates at 4–6 EBs/well with 1 mL NK medium (*see* **Note 6**).

3. Half-medium changes are done every 5–6 days. The first week the NK cell differentiation medium contains 10 ng/mL IL-3 but removed when the first medium change. Phenotyping of NK Cell Development by Flow Cytometry (*see* **Note 7**).
4. Mature CD45^{+}CD56^{+} NK cells can be phenotyped by flow cytometry. In vitro function of hES/iPS cell-derived NK cells can be analyzed by measurement of direct cytolytic activity tumor cells (such as K562) by a standard ^{51}Cr-release assay or immunological assays for cytotoxic granule or cytokine release [1–3] (*see* **Note 8**).

4 Notes

1. M210-B4 feeder layer should be prepared the day but no longer than 3 days prior to use. Also, after certain passages (>25), stromal cells will decrease their ability to support hematopoietic differentiation.
2. 1 mg/mL collagenase medium is made fresh. Do not store for longer than 2 weeks at 4 °C.
3. hESC/iPSC colonies should be starting to differentiate 3–4 days after they are transferred onto M210-B4 feeder layer. Colonies spread out from defined edges and a heterogeneous population of cells should appear.
4. 0.05 % Trypsin-EDTA+2 % chick serum should be made fresh. To make sure single suspension cells have good viability, don't leave cell cultures in trypsin longer than 10 min even though there still are cell clumps not completely digested.
5. EBs should have started to form after 24 h incubation. But if hESC/iPSCs have not been well adapted by TrypLE (at least >10 passages) or have been passed more than 35–40 passages, EBs may not be formed well.
6. EL08-1D2 stromal cells are a temperature sensitive cell line, which should be maintained at an incubator with 33 °C, 5 % CO_2. However, cells can be kept at 37 °C after inactivation.
7. Poor NK differentiation from hematopoietic progenitors on ELs. If the feeders have been passed in culture for more than 2 months (typically more than ten passages), they may not support NK cell development effectively. The resource, different lots of cytokines also affect NK cell differentiation. Also make sure cytokines are stored at −20 °C and no longer than a week at 4 °C. If the purity/viability of CD34^{+}CD45^{+} progenitors

isolated from M210-B4 s is poor, it also affects NK cell differentiation. Of course, if the hematopoietic differentiation from spin EBs does not undergo well, the NK cell differentiation efficiency will decrease as well.

8. It usually takes longer (about 5 weeks) to derive NK cells from spin EBs than from enriched progenitors from stromal coculture. It is likely the hematopoietic progenitors in d11 spin EBs are still in earlier development status compared with those from d21 M210-B4 cocultures.

References

1. Woll PS, Martin CH, Miller JS, Kaufman DS (2005) Human embryonic stem cell-derived NK cells acquire functional receptors and cytolytic activity. J Immunol 175:5095–5103
2. Woll PS, Grzywacz B, Tian X et al (2009) Human embryonic stem cells differentiate into a homogeneous population of natural killer cells with potent in vivo antitumor activity. Blood 113:6094–6101
3. Ni Z, Knorr DA, Clouser CL et al (2011) Human pluripotent stem cells produce natural killer cells that mediate anti-HIV-1 activity utilizing diverse cellular mechanisms. J Virol 85(1):43–50
4. Wilber A, Linehan JL, Tian X et al (2007) Efficient and stable transgene expression in human embryonic stem cells using transposon-mediated gene transfer (Dayton, Ohio). Stem Cells 25:2919–2927
5. Giudice A, Trounson A (2008) Genetic modification of human embryonic stem cells for derivation of target cells. Cell Stem Cell 2:422–433
6. Ng ES, Davis R, Stanley EG, Elefanty AG (2008) A protocol describing the use of a recombinant protein-based, animal product-free medium (APEL) for human embryonic stem cell differentiation as spin embryoid bodies. Nat Protoc 3:768–776
7. Ng ES, Davis RP, Azzola L, Stanley EG, Elefanty AG (2005) Forced aggregation of defined numbers of human embryonic stem cells into embryoid bodies fosters robust, reproducible hematopoietic differentiation. Blood 106:1601–1603
8. Kaufman DS, Hanson ET, Lewis RL, Auerbach R, Thomson JA (2001) Hematopoietic colony-forming cells derived from human embryonic stem cells. Proc Natl Acad Sci U S A 98:10716–10721
9. Ledran MH, Krassowska A, Armstrong L et al (2008) Efficient hematopoietic differentiation of human embryonic stem cells on stromal cells derived from hematopoietic niches. Cell Stem Cell 3:85–98
10. Tian X, Morris JK, Linehan JL, Kaufman DS (2004) Cytokine requirements differ for stroma and embryoid body-mediated hematopoiesis from human embryonic stem cells. Exp Hematol 32:1000–1009
11. Ng ES, Davis RP, Hatzistavrou T, Stanley EG, Elefanty AG (2008) Directed differentiation of human embryonic stem cells as spin embryoid bodies and a description of the hematopoietic blast colony forming assay. Curr Protoc Stem Cell Biol; Chapter 1:Unit 1D 3

Chapter 4

Evaluation of Immunogenicity of Rat ES-Cell Derived Endothelial Cells

Martina Seifert and Juliane Ladhoff

Abstract

Evaluation of the immunogenicity of embryonic stem cell derived differentiated cells is important for their potential application in cell replacement therapies and transplantations. Low immunogenicity or even an immune privileged status would enable their general use in allogeneic settings and therefore supply an unrestricted source. Based on conflicting data in terms of immunogenicity published for mouse and human ES-derived cells, the rat model was used to complement the knowledge in this specific area by a set of in vitro test systems using endothelial ES cell derivatives.

This chapter describes the strategies and methods used to analyze immunogenicity of rat ES cell derived endothelial cells (RESC) in comparison to adult mature rat endothelial cells (EC). In a first characterization step, the endothelial nature of rat ES cell derived endothelial cells was proved by labelling with von Willebrand factor (vWF) as well as testing tube formation capacity on an extracellular matrix. The RESC can be characterized by their constitutive or cytokine-induced expression level of the Major Histocompatibility Complex (MHC) antigens class I and class II by Fluorescence Activated Cell Sorter (FACS) technology. Moreover, regulation of transcription factors involved in the IFNγ signaling pathway could be evaluated by detecting either the phosphorylation status by specific intracellular antibody staining followed by flow cytometric measurement or by analyzing the mRNA expression level by quantitative RT-PCR. By stimulating the RESC with IFNγ and coculturing with Carboxy-fluorescein diacetate succinimidyl ester (CFDA-SE)-labelled $CD4^+$ rat T cells, the ability of RESC to induce proliferation was analyzed by FACS technology. Allo-reactive cytotoxic T cells were generated in a mixed lymphocyte culture (MLC) with lymph node cells from two MHC-disparate rat strains and used to determine the susceptibility of RESC to lytic processes. Therefore, RESC were labelled with calcein and the release of this fluorochrome after coculture was measured. To analyze the response to humoral attacks, RESC were incubated with allo-antibody containing sera and rabbit complement and then cell damage was assessed by 7-actinomycin D (7-AAD) incorporation into the DNA using FACS analysis.

Key words Stem cells, Cell differentiation, Endothelial cells, MHC antigens, Class II transactivator (CIITA), Immunogenicity, Proliferation, Cytotoxicity

Nicholas Zavazava (ed.), *Embryonic Stem Cell Immunobiology: Methods and Protocols*, Methods in Molecular Biology, vol. 1029, DOI 10.1007/978-1-62703-478-4_4, © Springer Science+Business Media New York 2013

1 Introduction

Embryonic stem cells (ESC) are considered to be powerful tools in regenerative and transplantation medicine. Besides their capacity to differentiate into several adult-type cell lineages, aspects of their immunogenic potential and their behavior under transplantation conditions require more attention. Descriptions by other groups vary between an immune-privileged state [1, 2] and profound immunogenicity following allogeneic transplantation with subsequent differentiation [3, 4]. In a rat model system we used a well-characterized rat ES cell-like cell line [5] and generated endothelial derivatives as described [6]. We have recently analyzed the complex pattern of adaptive immune response against rat ES-cell derived endothelial cells (RESC) in more detail [7]. In this context, the surface expression of structures recognized by the host immune system such as MHC class I and II play a predominant role in the recognition of grafted non-autologous (allogeneic) cells by the recipient's immune system [8].

This chapter describes the impact of MHC class II up-regulation under inflammatory conditions on the cellular and humoral immune responsiveness of RESCs in comparison to adult aortic rat endothelial cells (EC), their normal somatic counterpart. RESC derived cells showed impaired MHC class II expression after IFNγ stimulation, reduced allo-stimulatory capacity, diminished susceptibility to cytotoxic T cell lysis, and protection against humoral allo-recognition by allo-antibodies and complement. The overall reduced adaptive immune responses to RESC derivatives support the conclusion that these rat ES derived cells are immune privileged.

2 Materials

2.1 General Materials

1. Phosphate buffered saline (PBS; PAA, Pasching, Austria).
2. Dulbecco's modified Eagles medium (DMEM; PAA, Pasching, Austria).
3. RPMI 1640 (PAA, Pasching, Austria).
4. Endothelial Basal Medium (EBM; PAA, Pasching, Austria).
5. Penicillin 10.000 U/ml (Life Technologies, Karlsruhe, Germany).
6. Streptomycin 10.000 μg/ml (Life Technologies, Karlsruhe, Germany).
7. L-Glutamine 200 mM (Life Technologies, Karlsruhe, Germany).
8. Fetal calf serum (FCS; PAA, Pasching, Austria) (inactivated, *see* **Note 1**).

9. Nonessential amino acids 100× (Life Technologies, Karlsruhe, Germany).
10. β-Mercaptoethanol (Merck, Darmstadt, Germany): prepared as a 1 M stock solution in RPMI medium and stored at 4 °C.
11. Nucleosides (adenosine, guanosine, cytosine, uridine, thymidine) (Sigma, Steinheim, Germany): prepared as a nucleoside solution with 8 μg/ml adenosine, 8.5 μg/ml guanosine, 7.3 μg/ml cytosine, 7.3 g/ml uridine, 2.4 μg/ml thymidine by dissolving in distilled water at 37 °C. Sterile filtered and aliquoted solution could be stored at 4 °C.
12. Recombinant human insulin (Roche Diagnostics, Mannheim, Germany): prepared as a 10 mg/ml stock solution in distilled water and further diluted 1:2 with PBS/10 % (v/v) FCS and stored in aliquots at −20 °C.
13. Recombinant rat interferon γ (IFNγ; Hycult biotechnology B.V., Uden, Netherlands).
14. Trypsin/ethylenediamine tetraacetic acid (EDTA), 100× (Life Technologies, Karlsruhe, Germany).
15. Trypan blue, 0.4 % (w/v) solution (Sigma, Steinheim, Germany).
16. Gelatine (Sigma, Steinheim, Germany): prepared as a 2 % (w/v) solution in sterile distilled water heated to 65 °C, then filtered through a 0.22 μm sterile filter and stored in a glass bottle at 4 °C. For coating culture plastics, a working solution of 0.2 % (w/v) is prepared by 1:10 dilution of the 2 % stock solution in PBS.
17. 15 ml tubes (Falcon, Oxnard, USA).
18. 50 ml tubes (Falcon, Oxnard, USA).
19. Tissue culture flask T-25, 25 cm^2 (Falcon, Oxnard, USA).
20. Tissue culture flask T-75, 75 cm^2 (Falcon, Oxnard, USA).
21. 6-, 12-, 24-well plate (Falcon, Oxnard, USA).
22. 96-well plate U- and flat bottom plate (Nunc, Roskilde, Denmark).
23. 96-well plate flat bottom plate, black (Nunc, Roskilde, Denmark).
24. Cryovials (Nunc, Roskilde, Denmark).
25. Cell strainer (40, 100 μm) (Falcon, Oxnard, USA).
26. Syringe 1 or 2 ml (BD, Heidelberg, Germany).
27. 26 G$^{3/4}$ Needle, BD Microlance™ 3 (BD, Heidelberg, Germany).
28. Transfer pipettes (Sarstedt, Nümbrecht, Germany).
29. Petri dishes (BD, Heidelberg, Germany).
30. FACS tubes (Micronic Europe B.V., Lelystad, The Netherlands).
31. Eppendorf tubes (Eppendorf, Hamburg, Germany).

2.2 Cell Culture

1. Medium for RESC endothelial derivatives: DMEM supplemented with 15 % (v/v) FCS, 100 U/ml Penicillin, 100 μg/ml Streptomycin, 1 % (v/v) MEM nonessential amino acids, 1 % (v/v) nucleoside solution, 10 mM β-mercaptoethanol, and 1.7 nM human recombinant insulin.
2. Medium for primary rat aortic endothelial cells (EC): EBM supplemented with 10 % (v/v) FCS, 50 μg/ml Gentamycin (Sigma, Steinheim, Germany), 2 mM L-Glutamine and 6 μg/ml endothelial cell growth supplement (CellSystems, St. Katharinen, Germany).
3. Medium for mixed lymphocyte culture (MLC): RPMI 1640 supplemented with 10 % (v/v) FCS, 2 mM L-glutamine, 100 U/ml penicillin, 100 μg/ml streptomycin, 50 μM β-mercaptoethanol, 5 mM 2-(4-(2-hydroxyethyl)-1-piperazinyl)-ethansulfonsäure (HEPES; Biochrom AG, Berlin, Germany), and 2 % (v/v) autologous rat serum.

2.3 Tube Forming Assay

1. Matrigel™ (BD Biosciences Pharmingen, San Diego, USA).
2. RESC cell culture medium (*see* Subheading 2.2).
3. Inverse microscope Axio Observer Z1 (Carl Zeiss Microimaging GmbH, Göttingen, Germany) with AxioCam MRm system.
4. AxioVision Software; release 4.7.2 (Carl Zeiss Microimaging GmbH, Göttingen, Germany).

2.4 Cell Staining

1. FACS staining buffer: PBS with 2 % (v/v) FCS and 0.05 % (w/v) NaN_3.
2. BD Cytofix™ fixation buffer (BD Biosciences Pharmingen, San Diego, USA).
3. Methanol (Baker Analyzed®, Mallinckrodt Baker B.V., Deventer, The Netherlands): 90 % (v/v) stock solution.
4. Paraformaldehyde (PFA; Sigma, Steinheim, Germany): prepared a 1 % (w/v) solution in PBS (*see* **Note 2**).
5. T-Octylphenylpolyethylenglycol (Triton X®-100; 0.3% (w/v), Serva, Heidelberg, Germany) diluted in PBS.
6. Primary antibodies:
   - Anti-rat CD4-PE (BD Biosciences Pharmingen, San Diego, USA).
   - Anti-rat CD31 (PECAM-1; BD Biosciences Pharmingen, San Diego, USA).
   - Anti-rat CD54 (ICAM-1; BD Biosciences Pharmingen, San Diego, USA).
   - Anti-rat CD106 (VCAM-1; BD Biosciences Pharmingen, San Diego, USA).

   - IgG_1k isotype control (MOPC-21; BD Biosciences Pharmingen, San Diego, USA).
   - Rat-RT1A (OX18; BD Biosciences Pharmingen, San Diego, USA).
   - Rat-RT1B (OX6; BD Biosciences Pharmingen, San Diego, USA).
   - Anti-rat CD144 (VE-Cadherin; Alexis Biochemicals, Lausen, CH).
   - Rabbit isotype control serum (Sigma, Steinheim, Germany).
   - Anti-phospho STAT1-PE (pY701; part of the BD™ Phosflow kit; BD Biosciences Pharmingen, San Diego, USA).
   - Anti-von Willebrand factor (Sigma, Steinheim, Germany).

7. Secondary antibody conjugates:
   - PE-labelled donkey-anti-mouse IgG (H+L) polyclonal antibody (Dianova, Hamburg, Germany).
   - PE-labelled goat-anti-rabbit IgG (H+L) polyclonal antibody (Southern Biotechnology, Birmingham, USA).
8. BD FACSCalibur (Becton Dickinson, San Jose, USA).
9. CellQuest Software (Becton Dickinson, San Jose, USA).
10. Inverse microscope Axio Observer Z1 (Carl Zeiss Microimaging GmbH, Göttingen, Germany) with AxioCam MRm system.
11. AxioVision Software; release 4.7.2 (Carl Zeiss Microimaging GmbH, Göttingen, Germany).

2.5 Quantitative Real-Time Polymerase Chain Reaction

1. Absolutely RNA RT-PCR Miniprep Kit (Stratagene, La Jolla, USA).
2. qPCR-Mastermix (Eurogentec Deutschland GmbH, Köln, Germany).
3. SYBRGreen®-Mastermix (Applied Biosystems, Darmstadt, Germany).
4. RNA 6000 Reagents and Supplies (Agilent Technologies, Waldbronn, Germany).
5. dNTPs (Pharmacia Biotech, Uppsala, Sweden).
6. GoScript™ Reverse Transcription System (Promega GmbH, Mannheim, Germany).
7. DNAse I, RNAse free (Stratagene, La Jolla, USA).
8. RNAse inhibitor (RNAsin) (Promega GmbH, Mannheim, Germany).

9. RNA binding buffer (Promega GmbH, Mannheim, Germany).
10. Primer:

    Primer Rat Beta actin:

    5′-Primer: 5′-GTA CAA CCT CCT TGC AGC TCC T-3′.

    3′-Primer: 5′-TTG TCG TCG ACG ACG AGC GC-3′.

    Primer Rat IFNγ receptor chain 1:

    5′-Primer: 5′-AGT CGT CTT TCT GGC AAG TTA ATA CA-3′.

    3′-Primer: 5′-AAG CAG CAT CCA AAT TGA TTC TTC-3′.

    Primer Rat IFNγ receptor chain 2:

    5′-Primer: 5′-CAT CGC AGA GAC GAA ATG TGA-3′.

    3′-Primer: 5′-GCG CAG GAA GAC TGT GTA TGA GT-3′.

    Primer Rat class II transactivator:

    5′-Primer: 5′-CAA GGA CCT CTT CAT ACA GCA CAT T-3′.

    3′-Primer: 5′-GGA GGC ACT AGT TTC CTG TGC TT-3′.

    Primer Rat TNF intron:

    5′-Primer: 5′-TGA GAG AGT CAG AGC GGT GAT TC-3′.

    3′-Primer: 5′-CCT GCG CCC TCT GCT CTT-3′.

    Probe: 5′-ACG TCC CAT TGG CTA CGA GGT CCG-3′.
11. Agilent 2100 Bioanalyzer (Agilent Technologies, Waldbronn, Germany).
12. 7500 Real-Time PCR System (Applied Biosystems, Foster City, USA).
13. 7500 System SDS Software (Applied Biosystems, Foster City, USA).
14. MicroAmp Optical tubes and cups (Applied Biosystems, Foster City, USA).

2.6 Functional Analyses

2.6.1 Cytotoxicity Assay

1. Calcein-AM (MoBiTec, Göttingen, Germany): prepared a 1 mM stock solution in DMSO (Sigma Steinheim, Germany) and stored in aliquots at −20 °C (*see* **Note 3**).
2. RPMI 1640 without serum.
3. T-Octylphenylpolyethylenglycol (Triton X®-100; 1.8 % (w/v), Serva, Heidelberg, Germany) diluted in PBS.
4. 96-Well plate flat bottom plate, black (Nunc, Roskilde, Denmark).
5. Fluorescence microscope (Carl Zeiss Microimaging GmbH, Göttingen, Germany).
6. GENios Microplate reader (TECAN, Crailsheim, Germany).

2.6.2 Allo-Antibody/ Complement Assay

1. PBS with 2 % (v/v) FCS.
2. GVB^{++} buffer (Sigma, Steinheim, Germany).
3. Allo-antigen specific serum (*see* **Note 4**).
4. Rabbit complement (Accurate Chemicals, Westbury, USA): prepared as a 1 mg/ml stock solution in cold GVB^{++} buffer and stored in aliquots at −80 °C (*see* **Note 5**).
5. 7-Amino-actinomycin D (7-AAD; Sigma, Steinheim, Germany): prepared a stock solution of 1 mg/ml by dissolving 1 mg in 50 μl methanol (100 %) and adding 950 μl PBS before storing in aliquots at −20 °C. Working solution was diluted with PBS/2 % (v/v) FCS to 2.5 μg/ml.

2.6.3 One Way Mixed Lymphocyte Culture

1. MACS buffer: PBS with 0.5 % (w/v) bovine serum albumin (Sigma, Steinheim, Germany) and 2 mM EDTA (Merck, Darmstadt, Germany).
2. Rat CD4-MicroBeads (Miltenyi Biotec GmbH, Bergisch-Gladbach, Germany).
3. Medium for mixed lymphocyte culture (MLC): RPMI 1640 supplemented with 10 % (v/v) FCS, 2 mM L-glutamine, 100 U/ml penicillin, 100 μg/ml streptomycin, 50 μM β-mercaptoethanol, 5 mM HEPES and 2 % (v/v) autologous rat serum (heat inactivated at 56 °C for 30 min).
4. Carboxy-fluorescein diacetate succinimidyl ester (CFDA-SE; Molecular Probes, Oregon, USA): prepared a 5 mM stock solution in DMSO and stored in aliquots at −20 °C in the dark (*see* **Note 6**).
5. Mouse anti-rat CD4-PE antibody (BD Biosciences Pharmingen, San Diego, USA).
6. Concanavalin A (ConA; Sigma, Steinheim, Germany): prepared a 1 mg/ml stock solution in PBS and stored in aliquots at −20 °C.
7. MS columns (Miltenyi Biotec GmbH, Bergisch-Gladbach, Germany).
8. MiniMACS separator (Miltenyi Biotec GmbH, Bergisch-Gladbach, Germany).
9. MACS MultiStand (Miltenyi Biotec GmbH, Bergisch-Gladbach, Germany).
10. Blood radiation unit OB29 (Steuerungstechnik & Strahlenschutz (STS) GmbH, Braunschweig, Germany).
11. FACSCalibur (Becton Dickinson, San Jose, USA).
12. CellQuest software (Becton Dickinson, San Jose, USA).

3 Methods

All analyses were performed in comparison to primary rat aortic endothelial cells (EC) as counterparts isolated from adult differentiated tissue.

3.1 Cell Culture of RESC Endothelial Derivatives

RESC derivatives were differentiated from the RESC line C12 as described [7]. They were cultured on gelatine-coated dishes for all experiments.

3.1.1 Gelatine Coating of Cell Culture Dishes

1. Coat cell culture dishes with 0.1 % (w/v) gelatine.
2. Incubate for 1 h at room temperature.
3. Wash three times with PBS.
4. Immediately use the coated dishes or store them covered with serum free medium at 37 °C with a 5 % (v/v) CO_2 humidified atmosphere.

3.1.2 Cell Culture of RESC Endothelial Derivatives

1. Culture the RESC endothelial derivatives ensuring they do not reach confluence.
2. Change medium completely every 2–3 days.
3. Split the culture in a 1:3 ratio and passage (*see* Subheading 3.1.3) into fresh gelatine-coated culture dishes.

3.1.3 Cell Passage

1. Aspirate medium carefully.
2. Rinse once with PBS and aspirate again.
3. Add Trypsin/EDTA (2 ml for a 25 cm^2 culture flask or 5 ml for a 75 cm^2 flask) and incubate at 37 °C for 3–5 min until cells detach (*see* **Note 7**).
4. Stop the trypsinization process by adding 5 ml of culture medium.
5. Transfer cell suspension into a 15 ml tube.
6. Centrifuge at 200 × *g* for 10 min at 4 °C.
7. Discard the supernatant and resuspend the cells in culture medium.
8. Plate cells into gelatine-coated dishes.

3.2 Characterization of Endothelial Phenotype

3.2.1 Staining for von Willebrand Factor (vWF)

1. Culture the RESC derivatives in 24-well plates until reaching 80 % confluence and typical cobble stone morphology (Fig. 1a, d).
2. Aspirate the medium of each well carefully without disturbing the cell layer.
3. Wash wells three times with PBS by aspiration after carefully adding the buffer at the rim of the well.
4. Add 200 μl/well 4 % (v/v) PFA solution and incubate for 15 min at room temperature.

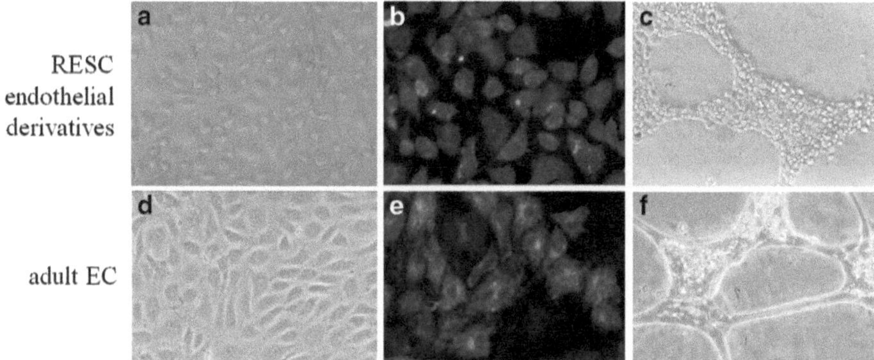

Fig. 1 Endothelial phenotype of RESC endothelial derivatives. Cobblestone-like morphology of RESC derivatives (**a**) and adult EC (**d**); intracellular von Willebrand factor expression (FITC-staining) in RESC endothelial derivatives (**b**) and adult EC (**e**); tube formation of RESC derivatives (**c**) and adult EC (**f**) on Matrigel™ after 24 h (200× magnification)

5. Aspirate the PFA solution and add 200 μl/well of 0.3 % (w/v) Triton®-X 100 solution followed by incubation for 30 min at room temperature to allow cells to become permeable.
6. Wash three times with PBS.
7. Add 200 μl of the polyclonal rabbit anti-vWF antibody in a 1:200 dilution in FACS-staining buffer or with the corresponding polyclonal rabbit serum as a control.
8. Incubate the cells in the culture well with the primary antibody solution 4 h at room temperature with continuous shaking.
9. Aspirate the primary antibody solution and add the FITC-labelled anti-rabbit-IgG secondary antibody (1:160 diluted in FACS staining buffer).
10. Incubate the secondary detection antibody 2 h at room temperature with continuous shaking.
11. Aspirate the secondary antibody solution and wash once with PBS.
12. Cover the wells with 200 μl PBS and store at 4 °C in a humidified box until use.
13. Examine the stained wells under a fluorescence microscope with 100× or 200× magnification. Excitation at 510 nm induces the FITC fluorescence (green emission).
    An example result is shown in Fig. 1b, e.

3.2.2 Tube Forming Assay

1. Prepare the Matrigel™ solution by slowly thawing it.
2. Pipet 250 μl Matrigel™ solution (9 mg/ml) immediately after it is thawed into wells of a 24-well plate. Avoid the generation of air bubbles (*see* **Note 8**).
3. Incubate the 24-well plate for 30 min at 37 °C in 5 % (v/v) CO_2 humidified air until the Matrigel™ becomes solid.

4. Harvest the RESC endothelial derivatives (*see* Subheading 3.1.3) and adjust the cell concentration to $5 \times 10^4/100$ µl RESC medium. Add 100 µl of cell suspension to each of the wells.
5. Incubate the seeded cultures for 30 min at 37 °C in 5 % CO_2 humidified air.
6. Add an additional 500 µl of RESC derivative medium to each of the 24 wells.
7. Incubate again at 37 °C in 5 % CO_2 humidified air until the first observation time point.
8. Document the tube formation under the inverse light microscope at 100–200× magnification after 4–6 h, then after 24 and 48 h. An example result after 24 h of culture on Matrigel™ is shown in Fig. 1c, f.

3.3 Cytokine Stimulation of RESC Endothelial Derivatives

1. Seed cells onto gelatine-coated cell culture plates to reach 70–80 % confluence at the end of the stimulation.
2. After 24 h change medium and add stimulation medium supplemented with 10 ng/ml IFNγ.
3. Incubate at 37 °C for 5 min, 6, 24, 48 or 96 h (depending on the following analysis).

3.4 FACS Staining of Extracellular Markers

1. Harvest the cells by trypsin treatment (*see* Subheading 3.1.3).
2. Centrifuge cell suspension at $200 \times g$ for 10 min at 4 °C.
3. Resuspend cell pellet in FACS staining buffer.
4. Add 5×10^5–1×10^6 cells/sample into FACS tubes.
5. Centrifuge at $200 \times g$ for 10 min at 4 °C and carefully aspirate supernatant.
6. Add 40 µl primary antibody solution in the appropriate concentration (*see* **Note 9**).
7. Briefly vortex and incubate for 30 min at 4 °C.
8. Wash with 1 ml of FACS-staining buffer.
9. Centrifuge at $200 \times g$ for 10 min at 4 °C and carefully aspirate supernatant.
10. Add 40 µl of secondary antibody solution.
11. Centrifuge at $200 \times g$ for 10 min at 4 °C and carefully aspirate supernatant.
12. Resuspend sample in 200–300 µl of FACS staining buffer (*see* **Note 10**).
13. Store samples at 4 °C until flow cytometric analysis.
14. Measure probes with the FACSCalibur and analyze using CellQuest software. Example results for the determination of MHC class I and II surface expression on RESC derived endothelial cells are shown in Fig. 2a, b, comparing unstimulated with IFNγ-stimulated cells.

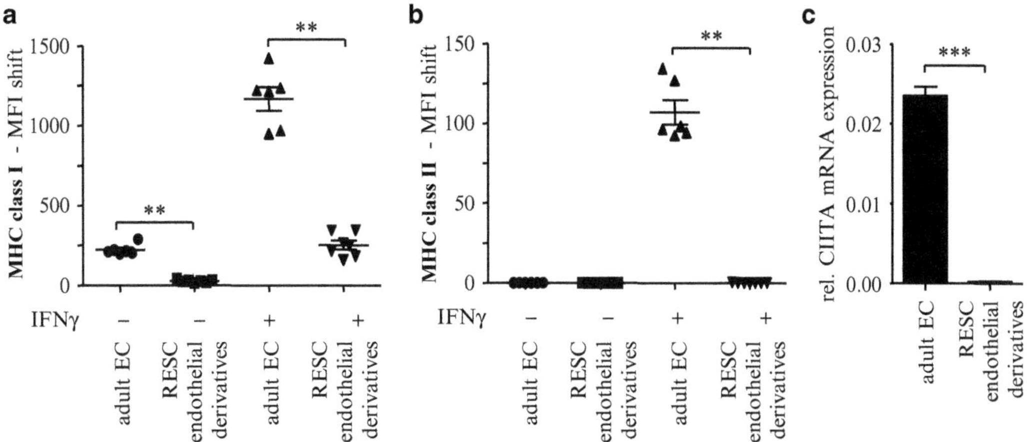

Fig. 2 Impaired cytokine mediated MHC upregulation by CIITA induction in RESC endothelial derivatives. (**a**) Reduced levels of basal and IFNγ-inducible surface MHC class I on RESC endothelial derivatives. (**b**) Lack of IFNγ-mediated surface MHC class II induction on RESC endothelial derivatives. Expression levels were analyzed by staining with mouse anti-rat-MHC I/MHC II specific antibodies followed by a PE-labelled anti-mouse-IgG secondary antibody and measured with flow cytometry after 24 h of cytokine stimulation. (**c**) Elimination of induction of class II transactivator (CIITA) mRNA following IFNγ treatment for 6 h, analyzed by qRT-PCR. Data presented as mean ± SEM of independent experiments ($n = 6$), **$p < 0.01$, ***$p < 0.001$

3.5 Analysis of Cytokine Induced STAT1 Phosphorylation

1. Culture the RESC endothelial derivatives in 24-well plates to 70 % confluence.
2. Stimulate the cells with stimulation medium for 5 min (*see* **Note 11**).
3. Aspirate medium and quickly rinse with PBS.
4. Harvest the cells by trypsin treatment (*see* Subheading 3.1.3; **Note 12**).
5. Transfer the cell suspension directly into FACS tubes.
6. Add 375 μl of BD Cytofix buffer to the samples and incubate for 10 min at 37 °C.
7. Centrifuge cell suspension at 200 × *g* for 10 min at 4 °C and aspirate supernatant.
8. Gently resuspend the cell pellet using a vortex.
9. Add 200 μl of 90 % methanol to each sample and incubate for 30 min on ice to make the cells permeable.
10. Centrifuge cell suspension at 200 × *g* for 10 min at 4 °C and aspirate supernatant.
11. Wash twice with 1 ml FACS staining buffer.
12. Centrifuge at 200 × *g* for 10 min at 4 °C and carefully aspirate supernatant.
13. Store samples at 4 °C until flow cytometric analysis.

3.6 Quantitative Real-Time PCR

3.6.1 mRNA Isolation and cDNA Synthesis

1. Isolate RNA from the cells using the "Absolutely RNA Miniprep Kit" according to the manufacturer's instructions.
2. Digest the RNA samples with DNAse for 15 min at 37 °C to eliminate DNA contamination.
3. Determine the RNA concentration of the samples, e.g., using the RNA 6000 Nano Assay Reagent & Supplies kit and the Agilent 2100 Bioanalyzer.
4. Store the samples until cDNA synthesis at −80 °C.
5. Use a maximum of 2 μg of RNA in a volume of 18 μl for reverse transcription into cDNA.
6. Add 2 μl oligodeoxythymidine (oligo dT) primer (0.5 μg/sample), to the RNA sample and incubate at 75 °C for 10 min to eliminate secondary structures.
7. Chill the samples for 2 min on ice.
8. Add 4 μl deoxynucleotides (0.5 mM each dNTP/sample), 2 μl DNase, 8 μl RNA binding buffer, and 0.5 μl RNase inhibitor (20 U/sample, RNasin) to each sample and incubate for 30 min at 37 °C.
9. Inactivate the DNase by incubating at 75 °C for 5 min.
10. Chill the samples for a further 2 min on ice.
11. Add 1 μl of reverse transcriptase (GoScript™) and 1 μl of RNase inhibitor to the samples and incubate for 60 min at 37 °C.
12. Stop the cDNA synthesis by incubating at 95 °C for 5 min.
13. Store cDNA at −20 °C until use.

3.6.2 Quantitative Real-Time PCR

1. Perform quantitative real-time PCR (qRT-PCR) analyses in a 96-well format in duplicates.
2. Use 1 μl of sample cDNA template for analysis.
3. Add SYBRGreen® Mastermix or qPCR Mastermix as directed, as well as the correct primer/probe mix to reach a total reaction volume of 13 μl.
4. Perform the qRT-PCR using the following conditions:

50 °C	2 min	Uracil-*N*-glycosylase (UNG) digest
95 °C	10 min	Inactivation of UNG, activation of AmpliTaqGold
95 °C	15 s	Denaturation
60 °C	1 min	Combined annealing and extension step (40 cycles in total)

5. For amplification use the 7500 Real-Time PCR System and analyze with the 7500 System SDS Software.
6. Normalize expression using the housekeeping gene β-actin.
7. When using SYBRGreen®, assess specificity of the desired gene products by melting curve analysis.
8. Calculate the relative expression levels of the target gene mRNA by means of the formula ($2^{-\Delta C_t}$). An example for the detection of mRNA expression levels for CIITA is shown in Fig. 2c.

3.7 One-Way Mixed Lymphocyte Culture

1. Plate stimulator cells (RESC endothelial derivatives) into a gelatine-coated 96-well plate and perform the IFNγ stimulation for 48 h as described above (*see* **Note 13**).

Isolation of lymph node cells

2. Harvest all lymph nodes from rats of the donor strain of the cells of interest and from a haplodiverse rat strain for comparison (e.g., Wistar-Kyoto and BDIX).
3. Transport the lymph nodes in PBS on ice.
4. Place a 100 μm cell strainer into a 10 cm dish and transfer the lymph nodes.
5. Mash the lymph nodes carefully through the cell strainer using a syringe plunger (*see* **Note 14**).
6. Rinse the cell strainer with 5 ml PBS.
7. Transfer the cell suspension into a 50 ml tube containing a 40 μm cell strainer.
8. Centrifuge cell suspension at 200 × *g* for 10 min at 4 °C and decant supernatant.
9. Wash once with 25–30 ml PBS.

Isolation of $CD4^+$ responder T cells

10. Wash lymph node cell suspension once with cold MACS buffer.
11. Determine cell number, using trypan blue to exclude dead cells.
12. Perform the $CD4^+$-MACS separation according to the manufacturer's instructions using the rat CD4 MicroBeads.

CFDA-SE labelling

13. Wash isolated $CD4^+$ T cells once with PBS.
14. Adjust the cell number to 1×10^7 cells/ml in PBS.
15. Add CFDA-SE to the cell suspension to reach a working concentration of 2.5 μM and incubate for 4 min at room temperature.
16. Stop the labelling by the addition of 15–20 ml MLC medium.

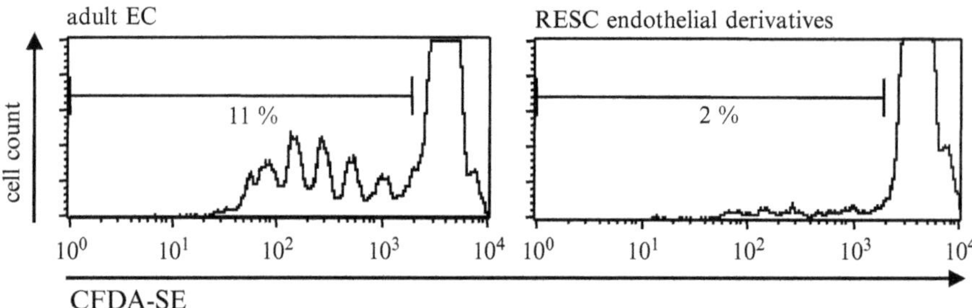

Fig. 3 Low allo-stimulatory capacity of RESC endothelial derivatives. Proliferation of allogeneic rat CD4$^+$ responder T cells in cocultures with IFNγ-pretreated adult EC and RESC endothelial derivatives on day 3 was analyzed by a CFDA-SE-based proliferation assay. Representative histograms of five independent experiments are shown

17. Wash three times with MLC medium. The cell pellet appears yellow/green in color during the washing steps.
18. Resuspend the labelled CD4$^+$ T cells in MLC medium supplemented with 2 % inactivated autologous serum.

Coculture

19. Irradiate the stimulator cells with 30 Gy and wash once with MLC medium.
20. Determine the actual cell number/well after the stimulation period by harvesting one well.
21. Add the CFDA-SE labelled CD4$^+$ responder cells to the stimulator cells applying a responder to stimulator ratio of 10:1 (*see* **Note 15**).
22. To assess autoproliferation, culture the isolated CD4$^+$ cells in MLC medium alone without stimulator cells.
23. Coculture the cells for up to 5 days at 37 °C in 5 % CO_2 humidified incubator.
24. Collect samples at days 3, 4 and 5 by carefully resuspending the proliferation clusters and transfer directly into FACS tubes.
25. Centrifuge samples at 200 ×*g* for 10 min at 4 °C and aspirate supernatant.
26. Wash once with FACS staining buffer.
27. Label the samples with an anti-rat CD4-PE antibody and subject them to flow cytometric analysis on a FACSCalibur using CellQuest software to gate on the CD4$^+$/CFDA-SE$^+$ population (*see* **Note 16**). Representative FACS histograms of proliferated CD4$^+$ T cells in coculture with the endothelial derivatives are shown in Fig. 3.
28. Calculate the stimulation index defined as the quotient of sample proliferation over the autoproliferation of CD4$^+$ cells.

3.8 Cytotoxicity Assay

Mixed lymphocyte culture

1. Harvest all lymph nodes from rats of the donor strain of the cells of interest and from a haplodiverse rat strain for comparison (e.g., Wistar-Kyoto and BDIX).
2. Transport the lymph nodes in PBS on ice.
3. Place a 100 μm cell strainer into a 10 cm dish and transfer the lymph nodes.
4. Mash the lymph nodes carefully through the cell strainer using a syringe plunger (*see* **Note 14**).
5. Rinse the cell strainer with 5 ml PBS.
6. Transfer the cell suspension into a 50 ml tube containing a 40 μm cell strainer.
7. Centrifuge the cell suspension at 200 × *g* for 10 min at 4 °C and decant supernatant.
8. Wash twice with 25–30 ml PBS.
9. Resuspend cell pellets in 20 ml precooled MLC medium and keep on ice.
10. Irradiate (donor derived) target lymphocytes with 30 Gy, keeping the tube on ice during irradiation.
11. Wash the target lymphocytes once with precooled (4 °C) MLC medium.
12. Count the cells using trypan blue in combination with 3 % (v/v) acetic acid to exclude erythrocytes. Alternatively, use Turk's solution for cell counting.
13. Plate the MLC into a 24-well plate with a responder: target ratio of 1:1 (3×10^6 responder cells and 3×10^6 target cells per well) in 2 ml MLC medium per well.
14. Culture MLC for 5 days at 37 °C in a 5 % CO_2 humidified incubator without medium exchange. Generated lymphoblasts are used as effector cells (cytotoxic T cells) in the calcein release assay.

Calcein release assay

15. Harvest the target cells (RESC endothelial derivatives) by trypsin treatment as described above.
16. Centrifuge cell suspension at 200 × *g* for 10 min at 4 °C and decant supernatant.
17. Wash cell suspension twice with PBS.
18. Count the cells using trypan blue to exclude dead cells.
19. Resuspend the cell pellet in serum-free medium for RESC endothelial derivatives with a cell concentration of $1–2 \times 10^6$ cells/ml.

20. Add Calcein-AM stock solution (1 mM) to the target cell suspension to reach the working concentration (20 μM) (*see* **Note 17**).
21. Gently shake the cell suspension and incubate for 30 min at 37 °C in the dark (e.g., inside an incubator). Gently shake once or twice during the incubation time.
22. Stop the cell labelling by the addition of 10 ml MLC medium.
23. Wash three times with MLC medium.
24. Count the cells under the fluorescence microscope to exclude unlabelled (nonfluorescent) cells (*see* **Note 18**).
25. Adjust the cell concentration on 2×10^5 target cells/ml in MLC medium.
26. During the calcein labelling: Harvest the effector cells generated from the 5-day MLC by carefully resuspending the proliferation clusters using a transfer pipette (*see* **Note 19**).
27. Collect effector cells in a 50 ml tube.
28. Centrifuge cell suspension at $200 \times g$ for 10 min at 4 °C and decant supernatant.
29. Wash once with MLC medium. Keep cells cooled on ice during the procedure.
30. Count the cells using trypan blue (*see* **Note 20**).
31. Perform a 1:2 serial dilution of the effector cells three times, starting with 2×10^6 effectors/100 μl (corresponding to an effector to target ratio of 100:1).
32. Pipette 100 μl of the calcein labelled target cell solution (2×10^5 target cells/ml) into a 96-well U-bottom plate (corresponding to 2×10^4 target cells/well). Use filter tips for pipetting. Perform triplicates or quadruplicates for each effector to target ratio (*see* **Note 21**).
33. In addition, prepare triplicates or quadruplicates for the estimation of the spontaneous calcein release and maximum calcein release.
34. Add 100 μl of the appropriate effector dilution to the target cells to reach the respective effector to target ratio.
35. To determine the spontaneous calcein release, add 100 μl of medium alone to the target cells.
36. To estimate the maximum calcein release, add 100 μl 1.8 % (w/v) Triton® X-100/PBS to the target cells.
37. Shortly centrifuge the plate for 30 s at $200 \times g$ to collect the cells at the bottom of the well.
38. Incubate the plate for 4 h at 37 °C in a 5 % CO_2 humidified incubator.

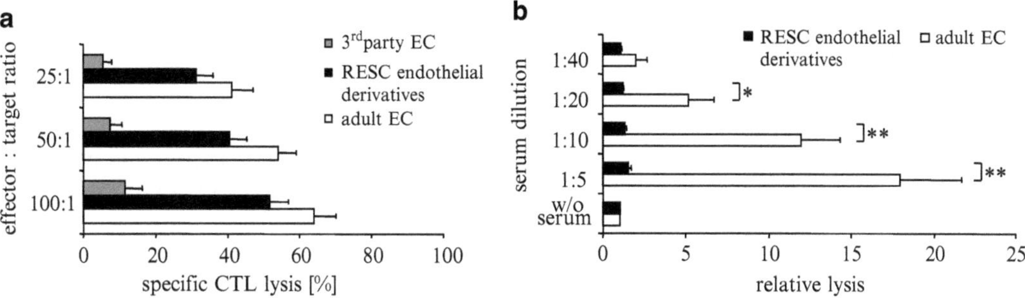

Fig. 4 Reduced immune responses in vitro. (**a**) Calcein-labelled RESC endothelial derivatives and adult EC were incubated with allo-specific cytotoxic T lymphocytes (CTLs). Allogeneic EC (*white bars*) were effectively killed by allo-specific CTL, whereas in RESC endothelial derivatives lysis was marginally reduced (*black bars*). Nonspecific killing was determined using adult third party EC (*grey bars*). Data are presented as mean ± SEM of independent experiments ($n=6$). (**b**) Allo-antibody/complement mediated lysis of RESC endothelial derivatives stayed at spontaneous lysis levels, whereas adult EC exhibited a 18-fold higher lysis rate compared to spontaneous lysis without serum incubation at a 1:5 dilution. Data are presented as mean ± SEM of independent experiments ($n=6$), $^{**}p<0.01$; $^{*}p<0.05$

39. Shortly centrifuge the plate again for 30 s at 200 × *g*.
40. Transfer 100 μl supernatant/well into a black 96-well flat bottom plate using filter tips (*see* **Note 22**).
41. Measure the fluorescence emission of the supernatants using the fluorimeter.
42. Calculate the specific lysis of the samples as follows:

$$\text{specific lysis}\,[\%] = \frac{(\text{experimental release} - \text{spontaneous release})}{(\text{maximum release} - \text{spontaneous release})} \times 100.$$

An example of calculated specific lysis rates is shown in Fig. 4a.

3.9 Allo-antibody/Complement Assay

1. Harvest the target cells (RESC endothelial derivatives) by trypsin treatment (*see* Subheading 3.1.3).
2. Wash once with PBS, centrifuge with 200 × *g* at 4 °C for 10 min.
3. Transfer 2×10^5 target cells into FACS tubes.
4. Centrifuge samples at 200 × *g* for 10 min at 4 °C and aspirate supernatant.
5. Perform dilutions of the allo-specific serum and the syngeneic control serum in PBS/2 % (v/v) FCS to gain appropriate dilutions (e.g., 1:5, 1:10, 1:20, 1:40).
6. Add 40 μl of the serum dilution to the corresponding sample. Include a negative control with PBS/2 % FCS without serum.
7. Incubate for 30 min at room temperature.

8. Stop the incubation by addition of 0.5–1 ml PBS/2 % FCS to each sample.
9. Centrifuge samples at 200×*g* for 10 min at 4 °C and aspirate supernatant.
10. Add 25 μl rabbit complement diluted 1:2 in ice cold GVB^{++} buffer to each sample and incubate for 1 h at 37 °C (*see* **Note 23**).
11. Dilute 7-AAD in PBS/2 % FCS to reach a working concentration of 2.5 μg/ml.
12. Add 200 μl of the 7-AAD containing solution/tube and incubate 15 min at room temperature in the dark.
13. Wash once with 1 ml PBS/2 % FCS. Centrifuge samples at 200×*g* for 10 min at 4 °C and aspirate supernatant.
14. Resuspend the cell pellet in 150–200 μl PBS/2 % FCS.
15. Perform flow cytometric analysis within the hour. Keep samples on ice until measurement.
16. Calculate relative lysis rates as the quotient of specific lysis$_{\text{7-AAD+ cells with allo-specific serum}}$/spontaneous lysis$_{\text{7-AAD+ cells without serum}}$.

An example result is shown for serum dilutions of 1:5–1:40 in Fig. 4b.

4 Notes

1. The batch of FCS used should be previously tested in cell cultures. Inactivate the FCS at 56 °C for 30 min and store the FCS in 50 ml aliquots at −20 °C.
2. Prepare a 5 % (v/v) stock solution in PBS by carefully heating to 55–60 °C on a hot plate in a fume hood to dissolve the PFA and then cool to room temperature before dilution with PBS for use.
3. Prepare aliquots with a small volume (e.g., 20 μl) to avoid repeated thawing and freezing. Prepare a working dilution of 20 μM by diluting 1:50 in PBS.
4. Allo-specific serum could be generated by the classical skin transplantation model between two rat strains with complete MHC mismatches such as WKY to BDIX using methods described elsewhere [9]. On the day of rejection, which depends on the combination of rat strains used, and in the case of WKY to BDIX occurs between days 10 and 12, blood from the skin graft recipient was collected by heart puncture and allowed to coagulate. Following storage for 24 h at 4 °C, serum was collected, centrifuged at 400×*g*, and carefully transferred to new tubes to avoid contamination by erythrocytes. Aliquots were prepared and then stored at −20 °C until

their use within the allo-antibody/complement assay. Allo-specificity was tested as described [10] by FACS analysis of responder rat thymocytes.

5. Baby rabbit complement should be preferentially used since other sources of complement do not work as well.

6. Prepare the working dilution of 2.5 μM CFDA-SE by adding 0.5 μl/1 ml of cell suspension in PBS. Working within a clean bench without the light on will reduce light damage to the CFDA-SE.

7. Control the trypsinization process under the inverse microscope. The cells become rounded as they detach and the flasks can be shaken and gently hit against a hard surface to help dislodge them for harvest.

8. Use bigger tips to prevent bubbles. If they occur, you can try to move them to the rim of the well with a 26 G¾ gauge needle.

9. Each antibody has to be titrated to find the optimal staining concentration for each cell type. For rat RESC derivatives and aortic endothelial cells, final concentrations of 5–10 μg/ml were commonly used.

10. If the stained FACS samples cannot be measured immediately, the cells can be fixed with 1 % (w/v) PFA. Samples are centrifuged after the last washing step at 200 × *g* for 10 min at 4 °C and the supernatants are aspirated. After careful resuspension of the pellet using a vortex, 100–200 μl of 1 % (w/v) PFA is added to each FACS tube and they are stored at 4 °C in the dark until analysis.

11. For each cell type the optimal time point to take the samples after stimulation has to be determined in a pre-experiment containing various time points.

12. At this step, trypsinization should be performed as quickly as possible.

13. To later determine the cell number of the adherent RESC endothelial derivatives in the coculture assay, two or three extra wells should be seeded. Before starting the coculture with the CFDA-SE-labelled $CD4^+$ T cells these wells are harvested and counted by trypan blue and the number of T cells to achieve a 1:10 ratio can be calculated.

14. The lymph nodes should be pressed very carefully to avoid cell damage.

15. To establish a positive control for the proliferation capacity of the rat $CD4^+$ T cells, it is recommended to add a well with ConA-stimulation (5 μg/ml) of the cells. When examining the proliferation during FACS analysis in a histogram plot, typical peaks for each division can be seen for the CFDA-SE-Channel (FL-1).

16. During each round of cell division the fluorescence signal of CFDA-SE is reduced by half, thus allowing the identification of successive cell generations. The probes are gated first for live cells (by forward and sideward scatter) and afterwards by further gating on all $CD4^+$ T cells. CFSE-staining is detected for all live and $CD4^+$ T cells by standard filters (492 nm excitation, 517 nm emission).
17. Use filter tips for this step to avoid contamination of your pipette with calcein.
18. Cell loss of up to 20 % of the initial cell number can be expected after finishing the calcein-labelling and subsequent washing steps. Therefore, it is recommended to label extra cells.
19. Use transfer pipettes for the harvest of the lymphoblastic cells and not normal pipette tips to avoid cell damage.
20. Count only the larger blast cells and not the smaller normal lymphocytes.
21. To avoid evaporation of the medium from the experimental wells, exterior wells of the 96-well plate should not be used as experimental wells and instead filled with PBS or medium.
22. Use a small needle to destroy bubbles that may interfere with the measuring process.
23. Each lot of complement should be tested before starting a key experiment. Try to order an appropriate number of complement vials to perform the complete series of experiments with the same lot.

Acknowledgments

The authors would like to thank Prof. Michael Bader for supplying the rat RESC cell lines. Thanks also to Elke Effenberger and Sabine Brösel for skillful technical assistance and Meaghan Stolk for manuscript correction. We also thank the BMBF for supporting the work within the Berlin-Brandenburg Center for Regenerative Therapies (BCRT).

References

1. Li L, Baroja ML, Majumdar A, Chadwick K, Rouleau A, Gallacher L, Ferber I, Lebkowski J, Martin T, Madrenas J, Bhatia M (2004) Human embryonic stem cells possess immune-privileged properties. Stem Cells 22:448–456
2. Bonde S, Zavazava N (2006) Immunogenicity and engraftment of mouse embryonic stem cells in allogeneic recipients. Stem Cells 24:2192–2201
3. Kofidis T, deBruin JL, Tanaka M, Zwierzchoniewska M, Weissman I, Fedoseyeva E, Haverich A, Robbins RC (2005) They are not stealthy in the heart: embryonic stem cells trigger cell infiltration, humoral and T-lymphocyte-based host immune response. Eur J Cardiothorac Surg 28:461–466
4. Swijnenburg RJ, Tanaka M, Vogel H, Baker J, Kofidis T, Gunawan F, Lebl DR, Caffarelli

AD, de Bruin JL, Fedoseyeva EV, Robbins RC (2005) Embryonic stem cell immunogenicity increases upon differentiation after transplantation into ischemic myocardium. Circulation 112:I166–I172

5. Fandrich F, Lin X, Chai GX, Schulze M, Ganten D, Bader M, Holle J, Huang DS, Parwaresch R, Zavazava N, Binas B (2002) Preimplantation-stage stem cells induce long-term allogeneic graft acceptance without supplementary host conditioning. Nat Med 8:171–178
6. Schulze M, Ungefroren H, Bader M, Fandrich F (2006) Derivation, maintenance, and characterization of rat embryonic stem cells in vitro. Methods Mol Biol 329:45–58
7. Ladhoff J, Bader M, Brosel S, Effenberger E, Westermann D, Volk HD, Seifert M (2009) Low immunogenicity of endothelial derivatives from rat embryonic stem cell-like cells. Cell Res 19:507–518
8. Chinen J, Buckley RH (2010) Transplantation immunology: solid organ and bone marrow. J Allergy Clin Immunol 125:S324–S335
9. Otomo N, Margenthaler JA, Motoyama K, Arima T, Shimizu Y, Lehmann M, Flye MW (2001) Organ transplant specificity of tolerance to skin grafts with heart or kidney grafts plus nondepleting anti-CD4 monoclonal antibody (RIB 5/2) and intravenous donor alloantigen administration. J Surg Res 98: 59–65
10. Ladhoff J, Fleischer B, Hara Y, Volk HD, Seifert M (2010) Immune privilege of endothelial cells differentiated from endothelial progenitor cells. Cardiovasc Res 88:121–129

Chapter 5

Interaction of ES Cell Derived Neural Progenitor Cells with Natural Killer Cells and Cytotoxic T Cells

Casimir de Rham and Jean Villard

Abstract

Knowing that human embryonic stem cells (HESC) can be derived into several different cells types render these cells very attractive to cure diseases. Unless these stem cells are originated from the patient itself, they will be isolated from a donor, who is genetically unrelated to the recipient. This situation will mimic an allogenic transplantation with an immune response against the transplanted cells. The immunogenicity of the HESC and the potential of NK and T-cells to target HESC and the lineage derived from HESC have to be addressed. Several different tests do exist to analyse NK cells and T-cells activity against HESC and its progenitor cells. In this chapter review the capacity of NK and T cells against neural progenitor derived from HESC, through a classical and a novel approach that combined the phenotype and also the functionality of the effector cells. In addition, we also demonstrate in the same test that we can determine the lysis of the progenitor cells by flow cytometry.

Abbreviations

7-AAD	7-Aminoactinomycin
CFSE	5-Carboxyfluorescein diacetate succinimidyl ester
FACS	Flow cytometry
HESC	Human embryonic stem cells
NK cells	Natural killer cells
NPC	Neuronal progenitor cells

1 Introduction

HESC are pluripotent stem cells, isolated from the inner cell mass of a human blastocyst [1]. Stem cells are cells found in a multicellular organism, and what renders them so interesting, is their ability to renew themselves. They are able to develop into the three primary germ layers: endoderm, mesoderm, and ectoderm [2]. Moreover, it is possible to derive these HESC into different progenitor cells, which can be maturated into several different cells

Nicholas Zavazava (ed.), *Embryonic Stem Cell Immunobiology: Methods and Protocols*, Methods in Molecular Biology, vol. 1029, DOI 10.1007/978-1-62703-478-4_5, © Springer Science+Business Media New York 2013

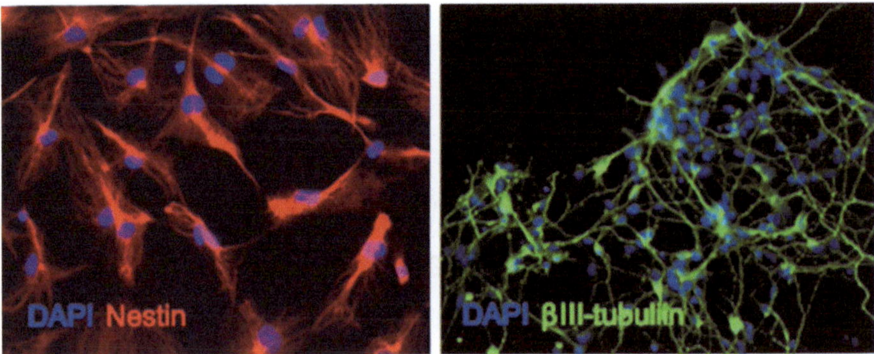

Fig. 1 *Left side*: Picture of NPC after 3 weeks of derivation from HESC. *Right side*: Picture of neurons after 2 weeks of derivation from NPC. From ref. 8

types (Fig. 1). The main goal of HESC is the transplantation of their progenitor cells to cure illness, or to reconstitute failing organs. Transplantation of NPC in the central nervous system can cure spinal cord injury, as shown in animal model [3]. But as the NPC will be originated from an individual, who is genetically unrelated to the receiver, a major problem arises. The host could react to the transplanted NPC and launch an immune response, in which NK and T-cells will be the main actors [4]. NK cells are part of the innate immune system and use a wide variety of inhibitory and activating receptors to regulate its activity [5]. NK cells do not need a maturation process to become effective as T-cells. They are able to kill cells, which express low or no self-MHC-I [6]. To determine T or NK cells activity, several experiments can be done. The chromium release assay is a usual test to determine the capacity of the effector cells to kill target cells. It is a simple test but with several inconvenient. This test uses a radioactive product, ^{51}Chromium, which means several precautions must be taken. On the other hand, nothing is known about the effector cells phenotype and functionality during this test. That is why a new test, using the flow cytometry, was set up. In one test, the phenotype and the functionality (cytotoxic assay, CD107a expression and IFN-γ secretion) of the effector cells can be analyzed [7, 8]. Briefly, target cells are stained with CFSE; this allows distinguishing them from the effector cells. After the cytotoxic assay, all the cells are harvest and 7-AAD is added just before the FACS analysis. 7-AAD is a staining for dead cells, this means that the double positive CFSE$^+$/7-AAD$^+$ population represent the target cells killed by the effector cells. So, on one side, target cells can be analyzed by gating on the CFSE$^+$/7-AAD$^+$ population, and on the other side, by gating on the effector cells, their phenotype and functionality can be studied.

2 Materials

2.1 Cells Isolation

1. Buffy Coat from a healthy donor, provided by the Blood Bank of the University Hospital Geneva (Geneva, Switzerland).
2. Ficoll-Paque™ Plus was from Amersham Biosciences (Uppsala, Sweden), stored at 4 °C.
3. Phosphate-buffered saline (PBS) by Gibco, Invitrogen (San Diego, CA, USA), stored at 4 °C.
4. Trypan Blue provided by Gibco, Invitrogen (San Diego, CA, USA), stored at room temperature.
5. Counting chamber (hemocytometer).
6. Human NK isolation KIT from Miltenyi Biotec (Bergisch Gladbach, Germany), stored at 4 °C.
7. Purified human anti-CD4 and purified human anti-CD14 are from Dako, Denmark. Purified human anti-CD19 and purified human anti-CD16 are from BD Biosciences, CA, USA. All stored at 4 °C.
8. Dynabeads® Pan Mouse IgG and magnet from Invitrogen (Oslo, Norway), stored at 4 °C.

2.2 Cells Culture

1. RPMI-1640 medium, from Gibco, Invitrogen (San Diego, CA, USA), stored at 4 °C.
2. Penicillin/streptomycin from Gibco, Invitrogen (San Diego, CA, USA), stored at −18 °C.
3. L-Glutamine from Gibco, Invitrogen (San Diego, CA, USA), stored at −18 °C.
4. Sodium pyruvate, Invitrogen (San Diego, CA, USA), stored at 4 °C.
5. Nonessential amino acids, Invitrogen (San Diego, CA, USA), stored at 4 °C.
6. β-Mercaptoethanol from Sigma Chemicals (St. Louis, MO, USA), stored at room temperature. Toxic product. Use standard laboratory procedures with toxic products.
7. Human serum provided by the Blood Bank of the University Hospital Geneva (Geneva, Switzerland), stored at −18 °C.
8. Fetal Calf Serum (FCS) from Sigma Chemicals (St. Louis, MO, USA), stored at −18 °C.
9. Stericup® 0.22 μm GP Millipore Express® PLUS membrane, by Millipore, MA, USA.
10. Human recombinant rhIL-15 provided by R&D System (Minneapolis, MN, USA). Stored at −80 °C in 10 μg/ml aliquots.

11. Human recombinant rhIL-2 provided by Biogen Inc. (Cambridge, MA, USA). Stored at −80 °C in 10 μg/ml aliquots.

2.3 CFSE Staining

1. 5-Carboxyfluorescein Diacetate Succinimidyl Ester (CFSE) from Molecular Probe, Inc. (Portland, OR, USA). Stored at −18 °C, stock at 5 mM. Light sensitive.
2. Phosphate-buffered saline (PBS) by Gibco, Invitrogen (San Diego, CA, USA), stores at 4 °C.
3. Fetal Calf Serum (FCS) from Sigma Chemicals (St. Louis, MO, USA) stored at 4 °C.
4. Trypsin/EDTA from Gibco, Invitrogen (San Diego, CA, USA), stored at −18 °C.

2.4 Functionality Assay

2.4.1 Chromium Release Assay (See Note 1)

1. Radioactive Sodium Chromate $Na_2{}^{51}CrO_4$ from Hartmann Analytics (Braunschweig, Germany) stored at 4 °C.
2. Phosphate-buffered saline (PBS) by Gibco, Invitrogen (San Diego, CA, USA), stores at 4 °C.
3. Triton X-100 from Sigma Chemicals (St. Louis, MO, USA) stored at room temperature. Irritant product. Use standard laboratory procedures with toxic products.
4. Macrowell™ tube strings by Skatron (Norway).
5. Radioactive γ-counter, PerkinElmer™ (MA, USA).

2.4.2 CD107 Expression

1. PE-conjugated human anti-CD107a from BD Biosciences (CA, USA), stored at 4 °C.
2. Monensin from Sigma Chemicals (St. Louis, MO, USA), stock at 2 mM Store at −18 °C. Toxic product. Use standard laboratory procedures with toxic products.

2.4.3 IFN-γ Secretion

1. Human IFN-γ secretion assay detection kit (APC) from Miltenyi Biotec (Bergisch Gladbach, Germany), stored at 4 °C.
2. Phosphate-buffered saline (PBS) by Gibco, Invitrogen (San Diego, CA, USA), stored at 4 °C.

2.5 Flow Cytometry Analysis

1. FACS buffer (PBS with 2 % FCS), stored at 4 °C.
2. PeCy7-conjugated human anti-CD56 from BD Biosciences (CA, USA), stored at 4 °C.
3. 7-Aminoactinomycin (7-AAD) from Sigma-Aldrich, Germany, stored at 4 °C, 1 mg/ml aliquots. Toxic product. Use standard laboratory procedures with toxic products.
4. 5 ml Polystyrene round-bottom tube from BD Biosciences (CA, USA).
5. FACS Aria from BD Biosciences (CA, USA).

3 Methods

3.1 Time Course

Day 0: Isolation of lymphocyte.
Isolation and culture of NK cells.
Isolation and culture of $CD8^+$ cells.

Day 6: CFSE staining.

Day 7: Cytotoxic assay, cells staining, and FACS analysis.

3.2 Cell Media

- In a 500 ml RPMI-1640 medium bottle, add:

  1 % penicillin/streptomycin.

  1 % L-glutamine.

  1 % sodium pyruvate.

  1 % nonessential amino acids.

  2 μl β-mercaptoethanol.

  Add for:

  T-cells culture: 10 % FCS.

  NK cells culture: 10 % Human serum (*see* **Note 2**).

  Filter the solution with a Stericup® 0.22 μm GP Millipore Express® PLUSmembrane.

3.3 Lymphocytes Isolation

1. Collect the blood from the buffy coat (60 ml) in a flask and add 60 ml of PBS.
2. Prepare four tubes (50 ml) and add 15 ml of ficoll.
3. Carefully pour the blood on the ficoll and centrifuge for 20 min at 540×*g*, without the brake.
4. Collect the white ring, which contain the lymphocytes, and pour them in a new tube (50 ml), complete to 50 ml with PBS. Centrifuge for 10 min at 135×*g*, with the brake.
5. Discard the supernatant, resuspend the pellet with PBS, and complete to 50 ml with PBS. Centrifuge for 15 min at 85×*g*, without the brake in order to discard the platelets.
6. Discard the supernatant and resuspend the pellet with 10 ml NK cells culture media.
7. Count the cells.

3.4 NK Cells Isolation

1. In a tube (15 ml) harvest 200 mio lymphocytes and complete with PBS.
2. Wash the cells for 10 min, 135×*g*.
3. Discard the supernatant.
4. Follow the human NK isolation KIT (from Miltenyi Biotec) protocol in order to isolate the NK cells.

5. Wash the cells for 10 min, 135×*g*.
6. Count the cells.
7. Place the NK cells, between 500,000 and 1 mio/ml, in a 24 wells plate.
8. Add IL-15 at a final concentration of 25 ng/ml.

3.5 T-cells (CD8+) Isolation

1. In a tube (15 ml) harvest 50 mio lymphocytes and complete with PBS.
2. Wash the cells for 10 min, 135×*g*.
3. Discard the supernatant.
4. Resuspend the lymphocytes in 500 μl of PBS.
5. Add 40 μl of CD4, CD14 and CD19, and 16 μl of CD16.
6. Incubate for 30 min on a agitator at 4 °C.
7. Complete with PBS and wash the cells for 10 min, 1,200 rpm.
8. Discard the supernatant.
9. Resuspend the lymphocytes in 4,687 μl of PBS.
10. Add 312 μl of Pan mouse IgG.
11. Incubate for 30 min on an agitator at 4 °C.
12. Mix well and place the tube in the magnet for 1 min.
13. Discard gently the supernatant, which contain the CD8+ cells.
14. Complete with PBS and centrifuge for 10 min, 135×*g*.
15. Count the cells.
16. Place the CD8+ cells, between 500,000 and 1 mio/ml, in a 24 wells plate.
17. Add IL-2 at a final concentration of 25 ng/ml.

3.6 CFSE Staining

1. Harvest the NPC.
2. Complete with PBS and centrifuge for 10 min at 300×*g* (*see* **Note 3**).
3. Discard the supernatant and resuspend the pellet in 2 ml of PBS.
4. Count the NPC.
5. Put 200,000 NPC in a 24 well with NK cells media. This will be the CFSEneg control, add 10 % of human serum (*see* **Note 4**).
6. Add 1 mio NPC in a tube (50 ml) and complete to 15 ml.
7. In a second tube (50 ml) add 15 ml PBS and 6 μl of CFSE (Stock: 0.5 mM).
8. Mix gently the CFSE solution with the NPC solution.
9. Incubate on an agitator for 15 min at 37 °C.
10. Block the reaction with 3 ml FCS.

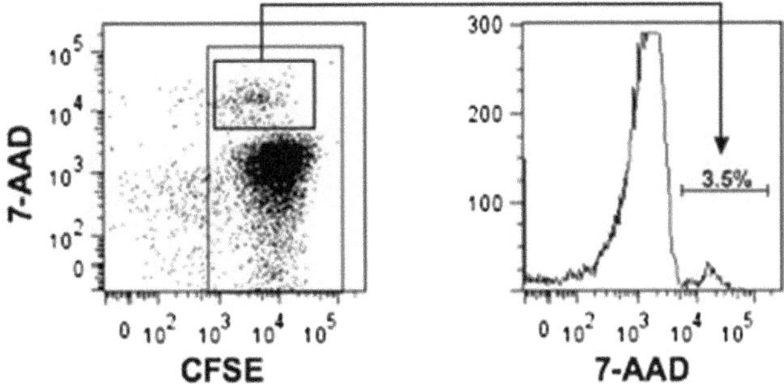

Fig. 2 Spontaneous lysis of NPC. The double positive CFSE+/7-AAD+ population represent the spontaneous dead NPC

11. Wash three times 10 min, 135×*g*, with PBS.
12. Count the NPC.
13. Put the NPC stained with CFSE in a flask with NK cells media and add 10 % human serum at 37 °C (*see* **Note 5**).

3.7 Cytotoxic Assay with Flow Cytometry

1. Harvest and count the effector cells (NK or CD8+ cells).
2. Harvest and count the target cells (NPC stained with CFSE).
3. Adjust the concentration of all cells at 50,000 cells/100 μl.
4. Put 50,000 target cells alone in a 96 wells plate U bottom. This will be the spontaneous lysis control (*see* **Note 6**) (Fig. 2).
5. Mix together 50,000 effector cells and 50,000 target cells in a 96 wells plate U bottom.
6. Incubate at 37 °C in a 5 % CO_2 humidified incubator.
7. After 1 h, add the PE-conjugated human anti-CD107a (3.5 μl), and monensin (3.5 μl from the 2 mM stock) [7].
8. Incubate the cells for 3 h more at 37 °C in a 5 % CO_2 humidified incubator.
9. Collect the cells and put them in a 96 well plate, V-bottom, for the IFN-γ secretion assay kit and for the FACS staining (Fig. 3).

3.8 ^{51}Chromium Staining

1. Incubate the NPC 1 h with $Na_2{}^{51}CrO_4$ at 37 °C.
2. Wash three times with PBS, in order to remove the ^{51}Chromium.
3. Control the level of γ-ray with the γ-ray counter (*see* **Note 7**).
4. Count the cells.

3.9 Cytotoxic Assay with ^{51}Chromium

1. In 12 wells of a 96 wells plate, U-bottom, put 10,000/100 μl target cells (NPC with ^{51}Chromium). Add 100 μl of media culture in the first six wells, and 100 μl of media culture with 10 % of triton-X100 in the last six wells (*see* **Note 8**).
2. In a 96 wells plate, U-bottom, mix 10,000/100 μl effector cells (NK cells) and 10,000/100 μl target cells (NPC with ^{51}Chromium).

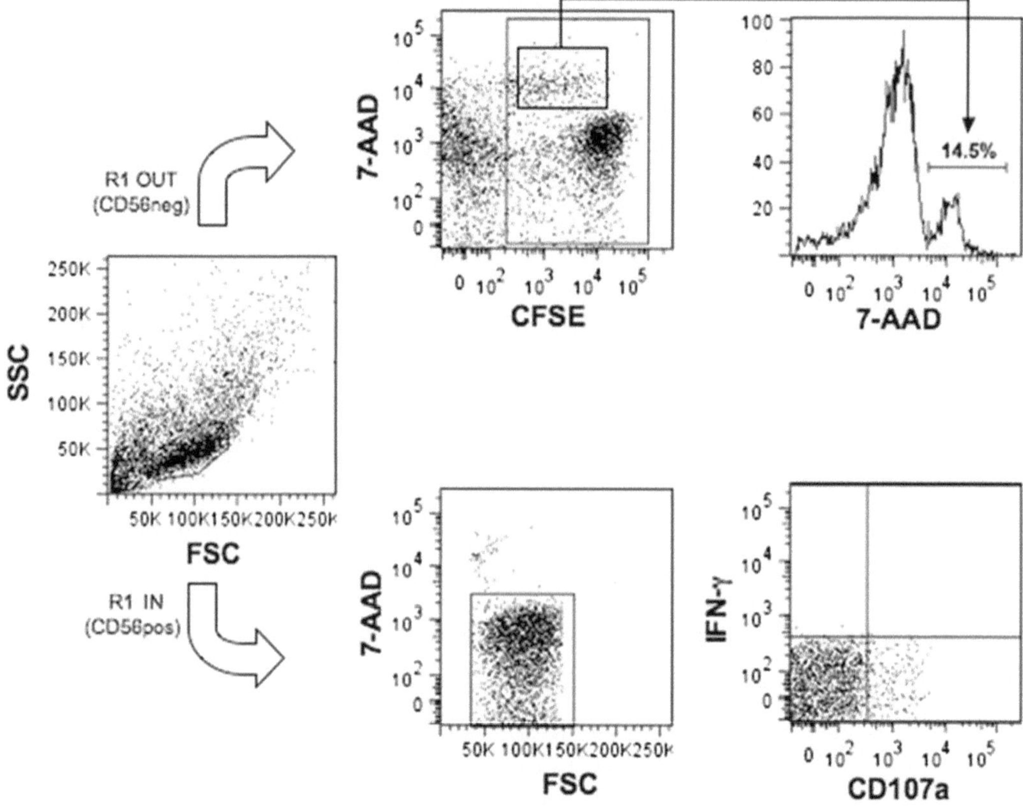

Fig. 3 Cytotoxic assay by FACS analysis. NK cells are gated and excluded from the analysis (*upper panel*), the double positive CFSE+/7-AAD+ NPC population represent NPC lysed by the NK cells. When the NK cells are gated on the living cells (*lower panel*), the expression of CD107a and the secretion of IFN-γ can be analyzed

3. Incubate for 4 h at 37 °C, in CO_2 humidified incubator.
4. Collect 100 ml of all the supernatant and put it in a Macrowell™ tube strings.
5. Put the Macrowell™ tube strings in a radioactive γ-counter.
6. The percentage of lysis can be calculated (*see* **Note 9**) (Fig. 4).

3.10 IFN-γ Secretion Assay and FACS Staining

1. Follow the Human IFN-γ secretion assay detection kit (APC) (from Miltenyi Biotec), protocol in order to detect IFN-γ secretion (*see* **Note 10**).
2. Wash the cells for 5 min at 1,200 rpm.
3. Discard the supernatant and add 25 μl of FACS buffer.
4. Stain the cells with PeCy7-conjugated human anti-CD56.
5. Incubate for 10 min on ice.
6. Wash the cells for 5 min at 1,200 rpm.
7. Prepare the FACS tube with 300 μl of FACS buffer.

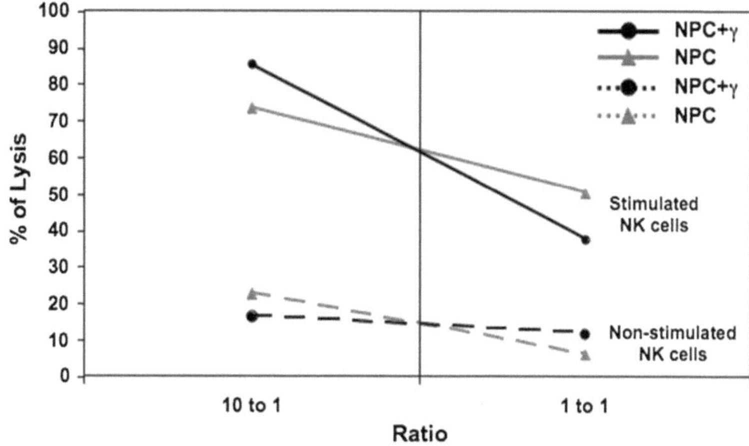

Fig. 4 ^{51}Cr release assay with NK cells stimulated or not with IL-15 as effector cells and NPC treated or not with IFN-γ as target cells. From ref. 8

8. Discard the supernatant, add 200 μl of FACS buffer, and put the cells in the FACS tube.
9. Add 3 μl of 7-AAD (Stock: 1 mg/ml) per tube and mix well (*see* **Note 11**).
10. FACS analysis.

4 Notes

1. The chromium release assay is a test using radioactive sodium chromate $Na_2{}^{51}CrO_4$. ^{51}Chromium produces γ-ray with a half-live of 27.7 days. All the procedures concerning working with radioactive products must be taken such as gloves, eyes protection and lab coat. Besides that, a special bench, designed for working with radioactive products, must be defined. This place will have a shield made of lead bricks, a γ-ray counter, and a waste for radioactive products. All the materials (pipettes, tips, PBS, etc.) used at this place must not be used elsewhere. All experiments containing ^{51}Cr must be done behind the lead shield.
2. It was observed that NK cells proliferate much better in presence of human serum comparing with FCS. Moreover, in presence of FCS, the mortality of NK cells increase.
3. As NPC are lighter than NK or T-cells, it is necessary to centrifuge them at a higher speed. This does not affect the NPC.
4. Once the number of NPC per milliliter in known, the great majority of them will be used for the CFSE staining. But a small number of the NPC need to be cultured in NK cells media as CFSE negative control for the FACS.

5. The CFSE stained or not NPC will stay overnight at 37 °C in a 5 % CO_2 humidified incubator. The next day, the supernatant is removing in order to get rid of the dead cells. The NPC are trypsinized and counted, the great majority will be used for the cytotoxic assay. A small number of the NPC need to be cultured in NK cells media as CFSE positive control for the FACS.
6. The spontaneous lysis will represent the lysis of the NPC without the presence of NK cells.
7. When the pellet, which has incorporated ^{51}Cr, shows 5 cps at the γ-ray counter, it can be used for the experiment. If this is not the case, one or two more centrifugations must be done.
8. In the first six wells the NPC with ^{51}Chromium will represent the spontaneous lysis. In the last six wells, due to the presence of triton X-100, which is a powerful detergent, the NPC will be totally lysed and all the ^{51}Chromium will be released in the supernatant. This will represent the maximal lysis. In the wells were NPC and NK cells are mixed, the NPC will released the ^{51}Chromium in the supernatant only if they are killed by the NK cells.
9. By using this formula:

$$\left[\frac{\text{nb of CPM} - \text{spontaneous lysis}}{\text{maximal lysis} - \text{spontaneous lysis}}\right] \times 100$$

   It will give the percentage of NPC, which are killed by the effector cells.
10. With this assay, we analyzed the cytotoxic assay by FACS. As FL-1 is used by the CFSE, FL-2 by CD107a PE, FL-3 by 7-AAD, the last empty channel will be the FL-4. That is why we used the IFN-γ secretion assay with an APC labeling.
11. The NPC stained with or without CFSE must be divided in two tubes. In each one of them the 7-AAD is added. This will allow having the four possible controls for the FACS. NPC with or without 7-AAD, and CFSE stained NPC with or without 7-AAD.

References

1. Thomson JA, Itskovitz-Eldor J, Shapiro SS, Waknitz MA, Swiergiel JJ, Marshall VS et al (1998) Embryonic stem cell lines derived from human blastocysts. Science 282(5391):1145–1147
2. Pera MF, Cooper S, Mills J, Parrington JM (1989) Isolation and characterization of a multipotent clone of human embryonal carcinoma cells. Differentiation 42(1):10–23
3. Morizane A, Li JY, Brundin P (2008) From bench to bed: the potential of stem cells for the treatment of Parkinson's disease. Cell Tissue Res 331(1):323–336
4. Cooper MA, Fehniger TA, Caligiuri MA (2001) The biology of human natural killer-cell subsets. Trends Immunol 22(11):633–640
5. Farag SS, Fehniger TA, Ruggeri L, Velardi A, Caligiuri MA (2002) Natural killer cell receptors: new biology and insights into the graft-versus-leukemia effect. Blood 100(6): 1935–1947

6. Ljunggren HG, Karre K (1990) In search of the 'missing self': MHC molecules and NK cell recognition. Immunol Today 11(7):237–244
7. Alter G, Malenfant JM, Altfeld M (2004) CD107a as a functional marker for the identification of natural killer cell activity. J Immunol Methods 294(1–2):15–22
8. Preynat-Seauve O, de Rham C, Tirefort D, Ferrari-Lacraz S, Krause KH, Villard J (2009) Neural progenitors derived from human embryonic stem cells are targeted by allogeneic T and natural killer cells. J Cell Mol Med 13: 3556–3569

Chapter 6

Strategies to Generate Induced Pluripotent Stem Cells

Michael Hayes and Nicholas Zavazava

Abstract

The isolation of embryonic stem cells (ESCs) has furthered our understanding of normal embryonic development and fueled the progression of stem cell derived therapies. However, the generation of ESCs requires the destruction of an embryo, making the use of these cells ethically controversial. In 2006 the Yamanaka group overcame this ethical controversy when they described a protocol whereby somatic cells could be dedifferentiated into a pluripotent state following the transduction of a four transcription factor cocktail. Following this initial study numerous groups have described protocols to generate induced pluripotent stem cells (iPSCs). These protocols have simplified the reprogramming strategy by employing polycistronic reprogramming cassettes and flanking such polycistronic cassettes with loxP or piggyBac recognition sequences. Thus, these strategies allow for excision of the entire transgene cassette, limiting the potential for the integration of exogenous transgenes to have detrimental effect. Others have prevented the potentially deleterious effects of integrative reprogramming strategies by using non-integrating adeno-viral vectors, traditional recombinant DNA transfection, transfection of minicircle DNA, or transfection of episomally maintained EBNA1/OriP plasmids. Interestingly, transfection of mRNA or miRNA has also been shown to be capable of reprogramming cells, and multiple groups have developed protocols using cell penetrating peptide tagged reprogramming factors to de-differentiate somatic cells in the absence of exogenous nucleic acid. Despite the numerous different reprogramming strategies that have been developed, the reprogramming process remains extremely inefficient. To overcome this inefficiency multiple groups have successfully used small molecules such as valproic acid, sodium butyrate, PD0325901, and others to generate iPSCs.

The fast paced field of cellular reprogramming has recently produced protocols to generate iPSCs using non integrative techniques with an ever improving efficiency. These recent developments have brought us one step closer to developing a safe and efficient method to reprogram cells for clinical use. However, a lot of work is still needed before iPSCs can be implemented in a clinical setting.

Key words Polycistronic casettes, piggyBac, Somatic cell nuclear transfer, cre/loxP

1 Introduction

The isolation of murine embryonic stem cells (ESCs) in 1981 and the subsequent isolation of human ESCs in 1998 has increased our understanding of normal embryonic development and opened the door for the development of stem cell derived treatments to debilitating diseases such as Type 1 diabetes and Parkinson's disease [1, 2].

Nicholas Zavazava (ed.), *Embryonic Stem Cell Immunobiology: Methods and Protocols*, Methods in Molecular Biology, vol. 1029, DOI 10.1007/978-1-62703-478-4_6,

However, the isolation of ESC requires the use of embryos, which is an ethically controversial approach. Therefore, new methods were needed to produce pluripotent stem cells without destroying embryos. This realization promoted the development of approaches involving the reprogramming of somatic cells. Although the field of cellular reprogramming is still in its infancy, the first successful reprogramming experiments took place nearly 60 years ago. The goal of these experiments, in which somatic cell nuclei from various amphibian species were transferred into enucleated eggs or zygotes, was not to reprogram cells but to determine if somatic cells possessed the full genetic complement of embryonic cells, as it had been previously hypothesized that as cells differentiated they would lose genetic material that was no longer needed. These first somatic cell nuclear transfer (SCNT) experiments were successful in demonstrating a somatic cell possesses a full genome and opened the door to the reprogramming of somatic cells to a pluripotent state [3].

Although it played a crucial role in our understanding of pluripotency and remains one of the most efficient methods available to reprogram somatic cells, SCNT is not an ideal strategy for producing pluripotent cells because it requires the use of an unfertilized egg (reviewed in Gurdon and Wilmut [4] and Gurdon et al. [5]). The isolation and manipulation of unfertilized eggs is commonly practiced in the setting of in vitro fertilization, but the scarcity of eggs in the research setting prohibits large scale use of SCNT. Thus, new technology is needed for the ethical and efficient generation of pluripotent stem cells.

In 2006 the Yamanaka group published a pioneering study in which they described reprogramming mouse cells with a defined cocktail of four transcription factors [6]. Soon after, both the Yamanaka and Thomson groups published studies describing the generation of human induced pluripotent cells (iPSCs) using a four transcription factor cocktail consisting of Oct4 (O), Sox2 (S), Klf4 (K), and cMyc (M) or Oct4, Sox2, Nanog (N), and Lin28 (L), respectively [7, 8]. The iPSCs generated by these methods have been demonstrated to be very similar to ESC (in morphology, pluripotency, mRNA profile, and protein expression) and many believe iPSCs will one day replace ESC as the pluripotent cell of choice for the development of clinical therapies.

Unlike ESCs, iPSCs generated using the retroviral transduction protocols described by the Yamanaka or Thomson groups contain transgenes that are known oncogenes (e.g., cMyc) [9]. These transgenes have been introduced via retroviral transduction and contain multiple viral integrations, thus introducing the possibility for dysregulation of tumor suppressors or proto-oncogenes [6, 10]. The integration of these oncogenic transgenes makes traditional iPSCs poor candidates for clinical use. For this reason, numerous groups have developed new methods to produce integration-free iPSCs. The reprogramming process is time

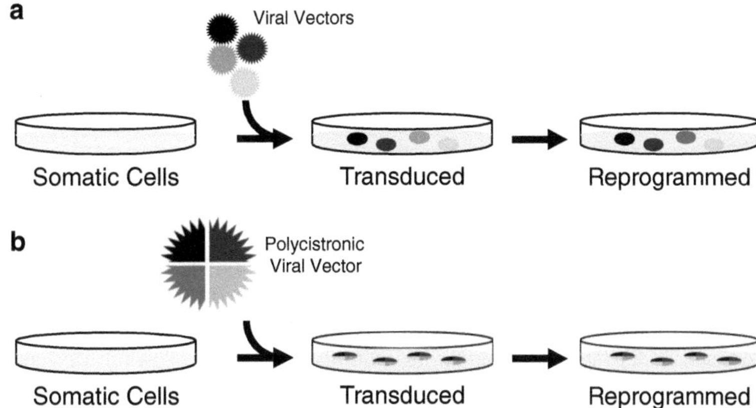

Fig. 1 Retroviral transduction. (**a**) The traditional approach to reprogram somatic cells with retroviral vectors involves the use of a cocktail of retroviruses each expressing a single reprogramming factor. Following viral transduction and transgene integration/expression somatic cells will dedifferentiate into iPSCs. iPSCs produced by this traditional multivector retroviral transduction contain multiple transgene integrations, potentially promoting neoplastic changes. (**b**) The development of polycistronic reprogramming vectors, where a single expression cassette contains multiple reprogramming factors linked by 2A sequences, reduced the number of viral integrations required to reprogram cells and thus decreased the risks associated with multiple transgene integrations

consuming, with traditional methods requiring 21–28 days. Further, the whole process is highly inefficient, as only 0.002 % of cells are successfully reprogrammed [6]. This review describes traditional reprogramming methods and those designed to decrease the introduction of exogenous genetic material and increase reprogramming efficiency.

2 Integrating Vectors

2.1 Retroviral Transduction

The first experiments to describe the generation of pluripotent stem cells without the need to sacrifice eggs or embryos employed retroviral transduction to deliver reprogramming factors [6–8]. Surprisingly only a four transcription factor cocktail is necessary to reprogram somatic cells. The efficiency of four factor reprogramming is low and has been shown to be improved with the inclusion of other factors such as Glis1, SV40LT, RARα, and LRH-1 [11–13]. Multiple retroviral vectors were originally needed to deliver the reprogramming cocktail (Fig. 1a). Now a single vector, employing picornaviral 2A sequences can deliver all of the reprogramming vectors using a single virus (Fig. 1b) [14]. This allows for fewer integrations and therefore reduces the likelihood of disturbing cellular homeostasis.

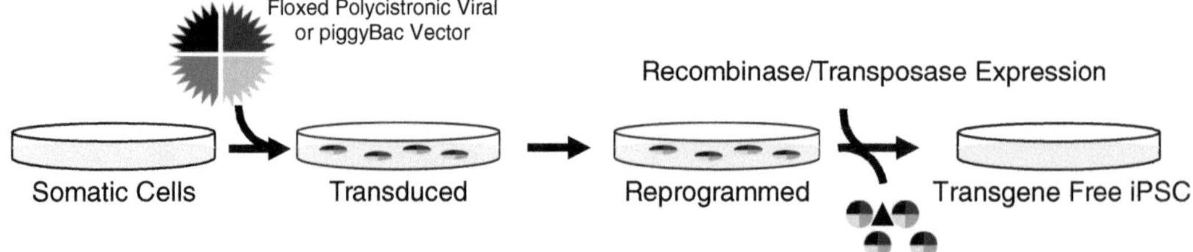

Fig. 2 Excisable reprogramming cassettes. Reprogramming using polycistronic retroviral transduction was further improved by flanking the reprogramming factor expression cassette with loxP sites. This floxed cassette can then be removed via transient expression of Cre recombinase. However, Cre mediated transgene excision is not ideal, as a single loxP site is left behind following transgene removal. The Woltjen group developed an alternative protocol, where piggyBac transposase is used to both integrate and seamlessly remove a polycistronic reprogramming cassette. The piggyBac protocol is advantageous in that the entire reprogramming cassette is removed without leaving a footprint, virtually eliminating the possibility of integration dependent alterations in gene expression

As described above, the initial reprogramming experiments used retroviral transduction to deliver reprogramming factors. Random integration of the reprogramming cassette can result in dysregulation of proto-oncogenes or aberrant expression of mutated protein sequences. Additionally, integrated transgenes may not be completely silenced, resulting in difficulty differentiating reprogrammed cells or an increased risk of developing neoplasia. These hurdles must be overcome before cells or tissues derived from reprogrammed somatic cells can be developed for clinical use. Therefore, an ideal integrative system would allow for the removal of transgenes post reprogramming.

2.2 Excisable Reprogramming Cassettes

The first methods used to excise integrated reprogramming cassettes take advantage of the Cre/LoxP system in the context of lentiviral transduction and integrative linear DNA transfection. Specifically, the integrated reprogramming cassette is flanked by loxP recombination sequences such that Cre expression in reprogrammed cells will excise the transgene cassette (Fig. 2) [14, 15]. Although these methods will successfully remove the reprogramming factors, excision using Cre/loxP leaves a single loxP site behind potentially disrupting endogenous protein expression. More recently the Woltjen group has developed a system employing the piggyBac transposase. Like Cre/loxP the piggyBac system effectively excises transgene sequences flanked by the piggyBac recognition sequences [15, 16]. However, unlike Cre/loxP and other transposases transposon excision utilizing piggyBac is clean because no exogenous DNA footprint is left behind.

The piggyBac system uses the piggyBac transposase to both introduce and remove a polycistronic reprogramming cassette. It is advantageous in that the reprogramming transgenes can be successfully removed without leaving a trace. However, reprogramming with piggyBac vectors requires an extra step to remove integrated transgenes, and extensive screening of iPSC colonies must be performed to ensure they are integration free. Even the most sensitive screening techniques may miss small single or oligomeric nucleic acid changes at excised integration sites [16]. Therefore the field has developed non-integrative methods to reprogram cells.

3 Non-integrating Vectors

The goal of developing iPSC-derived therapies has driven multiple groups to develop methods of generating iPSCs free of viral or transgene integrations as the integrations themselves and aberrant transgene expression are thought to be associated with an increased tumorigenicity [9]. Non-integrating reprogramming protocols are advantageous in that a transgene is never incorporated into a starting cell's genome, thus eliminating the risks associated with integrative approaches. These methods are diverse and each strategy has its own advantages and disadvantages. Although these techniques are attractive, they are also typically more tedious and can be less efficient than traditional integrative approaches.

3.1 Adenoviral Vectors

One of the first non-integrative approaches to produce both mouse and human iPSCs used adenoviruses encoding the Yamanaka reprogramming factors. Compared to their retroviral cousins, adenoviruses are non-integrating and therefore have a decreased risk of integration dependent alterations in cellular homeostasis.

Similar to retroviral-mediated reprogramming, adenoviral reprogramming requires transduction of a susceptible cell type with the four reprogramming factors Oct4, Sox2, Klf4, and cMyc (OSKM). Transduction of adenoviral encoded reprogramming factors will result in transient expression of these reprogramming factors, which is sufficient for dedifferentiation of somatic cells. Similar to piggyBac mediated reprogramming, characterization of adenoviral iPSCs will require screening (via Southern blot and PCR) to ensure clones are indeed transgene free [17, 18]. This is because adenoviruses are DNA viruses and therefore undergo integration events at a low frequency.

The attractiveness in adenoviral mediated reprogramming lies in the production of integration free iPSCs; however, the reprogramming efficiency is a dismal 0.0001–0.001 %. These experiments were completed in the absence of small molecules. Similar to retroviral reprogramming, adenoviral reprogramming efficiency may benefit from incorporation of one or more of the myriad small molecules available [17, 18].

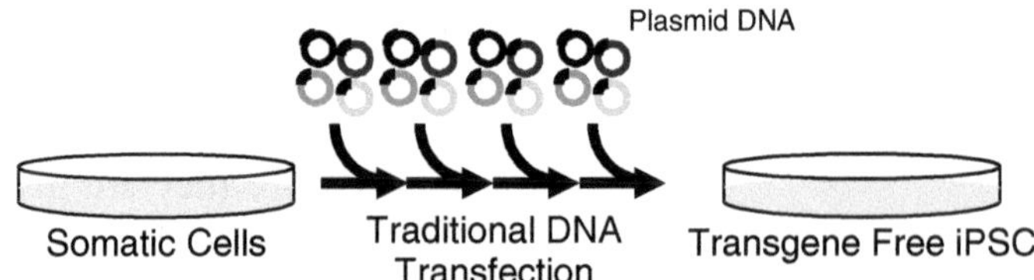

Fig. 3 Traditional DNA transfection. Transient transfection of recombinant DNA is a nonviral and non-integrating approach to reprogram somatic cells. This protocol, developed by Okita et al. requires multiple rounds of transfection owing to the short-lived nature of non-integrated DNA. Similar to other approaches, iPSCs generated using transiently transfected DNA will require intense screening to ensure no exogenous DNA sequences have become integrated

3.2 Traditional DNA Transfection

Another early integration free technique to generate iPSCs is the direct transfection of recombinant DNA, which is a virus free approach. Unlike the piggyBac and viral approaches described above, where a single treatment is sufficient for reprogramming, the transient nature of traditional DNA transfection necessitates multiple transfections to maintain the high protein expression necessary for reprogramming to occur (Fig. 3) [19, 20].

Similar to adenoviral transduction, transfection of recombinant DNA introduces the possibility of low frequency integration. PCR and Southern blot screening must be performed to ensure iPSC colonies are free of exogenous DNA. Unlike the adenoviral reprogramming methods, where no integration events were detected via PCR or Southern blot, multiple clones generated using this approach contained plasmid DNA sequences [19]. This relatively high rate of integration is concerning because that plasmid segments small enough to be below the limits of detection may be present in these iPSCs.

Reprogramming mouse embryonic fibroblasts using the traditional DNA transfection of four reprogramming factors is ~100–1,000-fold less efficient than traditional retroviral approaches. Similar to adenoviral mediated reprogramming, the incorporation of small molecules may increase the efficiency or decrease the number of transfections required to reprogram cells (also decreasing the probability of plasmid integration) [19].

The ease and availability of reagents for transient DNA transfection make this an attractive approach. However, the low reprogramming efficiency coupled with the requirement of multiple transfections and the relatively high rate of transgene integration make transient DNA transfection less than ideal. One way to improve upon this approach would be to increase the half-life of the transfected transgenes thereby removing the need for multiple transfections.

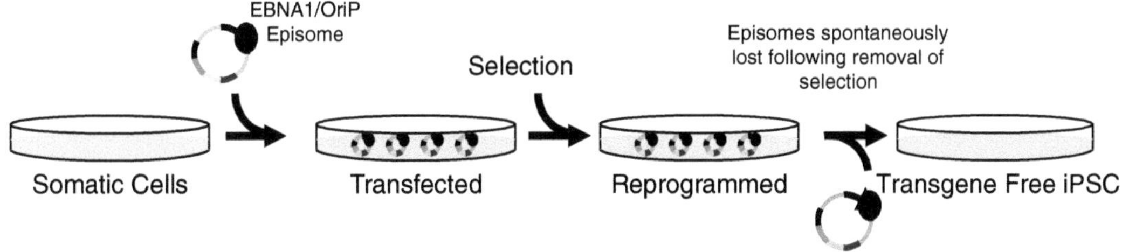

Fig. 4 EBNA1/OriP DNA. EBNA1 and OriP work together to ensure extrachromosomal replication and passage of OriP containing episomal vectors. Reprogramming vectors contain a selectable marker (e.g., drug resistance) such that episomes will be maintained when transfected cells are placed under selection. Upon removal of selection the reprogramming vectors are spontaneously lost, resulting in transgene free iPSCs

3.3 EBNA1/OriP DNA

Using a single DNA transfection the Thomson group was able to reprogram human fibroblasts utilizing the well characterized OriP/EBNA1 (Epstein–Barr nuclear antigen 1) system [21]. The Orip/EBNA1 system is derived from the Epstein–Barr virus and enables episomal vectors to be maintained indefinitely under drug selection. This is accomplished by incorporation of the cis-acting OriP element into a reprogramming vector containing a selectable marker and an EBNA1 expression cassette.

EBNA1 and OriP work together to ensure extrachromosomal replication and passage of the OriP containing episomal vectors (reviewed in Conese et al. [22]). Once selection is removed the episomal vectors are lost due to defects in replication and passage. Thus, transgene free iPSCs can easily be generated via single transfection of OriP/EBNA1 containing reprogramming vectors in selective media, with removal of reprogramming vectors following removal of drug selection (Fig. 4).

Reprogramming of human newborn fibroblasts with OriP/EBNA1 containing vectors, using a traditional four factor approach, is low at <0.01 %. However, OriP/EBNA1 mediated reprogramming efficiency can be increased by incorporating more reprogramming factors (e.g., SV40LT antigen) or by using a cell type that is more amenable to reprogramming such as cord blood and bone marrow derived mononuclear cells [21, 23]. EBNA1 expression may increase the immunogenicity of transfected cells but this has yet to be demonstrated in OriP/EBNA1 derived iPSCs and their derivatives.

3.4 Minicirle DNA

Minicircle DNA exists as small supercoiled DNA episomes consisting almost entirely of a eukaryotic protein expression cassette. Similar to the traditional DNA transfection and OriP/EBNA1 protocols described above, transfection of minicircle DNA can also be used to reprogram somatic cells [24]. The minimalistic nature of minicircle DNA (e.g., their lack of bacterial genes and prokaryotic origin of replication) results in decreased transcriptional silencing of prokaryotic sequences, culminating in an increased vector half-life and increased protein expression in eukaryotic cells.

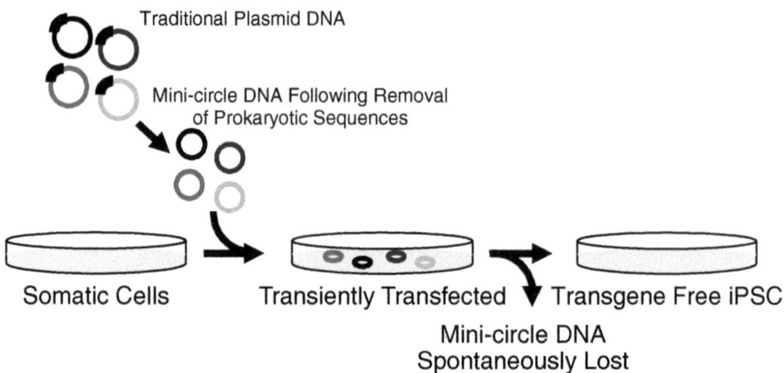

Fig. 5 Minicircle DNA. The transient nature of transgene expression following the transfection of traditional recombinant DNA construct can be partially attributed to the inclusion of prokaryotic sequences within these constructs. Minicircle DNA technology removes these prokaryotic sequences, thus increasing vector half-life necessitating only a single transfection. These vectors, which persist as episomes, are spontaneously lost resulting in the production of transgene free iPSCs

Jia et al. were able to successfully reprogram human adipose stromal cells (hASC, a cell type with endogenously high levels of Klf4 and cMyc) utilizing nulceofection followed by serial transfections of a single minicircle DNA vector carrying a polycistronic reprogramming cassette containing Oct4, Sox2, Lin28, Nanog, and GFP. 14–16 days after the initial transfection ESC like colonies began to form with a reprogramming efficiency of ~0.005 % (Fig. 5). Human newborn fibroblasts (HNF) were also reported to be reprogrammed to Tra-1-81$^+$ iPSC, but these iPSCs were not fully characterized [24].

Although minicircle mediated reprogramming holds promise for the easy production of transgene free iPSCs, there are multiple caveats to consider when using this approach. The production of minicircle DNA, as described, requires multiple culture conditions and special bacterial strains. Jia et al.'s protocol required multiple transfections and flow cytometric enrichment of transfected cells. Furthermore, all minicircle derived iPSC clones must be extensively screened to ensure exogenous DNA sequences have not been introduced. Further advances in minicircle technology, such as reprogramming with a greater number of factors or incorporation of small molecules into the protocol, may increase the efficiency and allow more cell types (e.g., adult human dermal fibroblasts) to be reprogrammed using these vectors.

4 Non-DNA Reprogramming

4.1 Modified mRNA

One non-integrative, non-DNA method used to successfully reprogram somatic cells moves down the central dogma of molecular biology and employs modified mRNA to express four reprogramming factors, therefore bypassing the need for DNA and avoiding

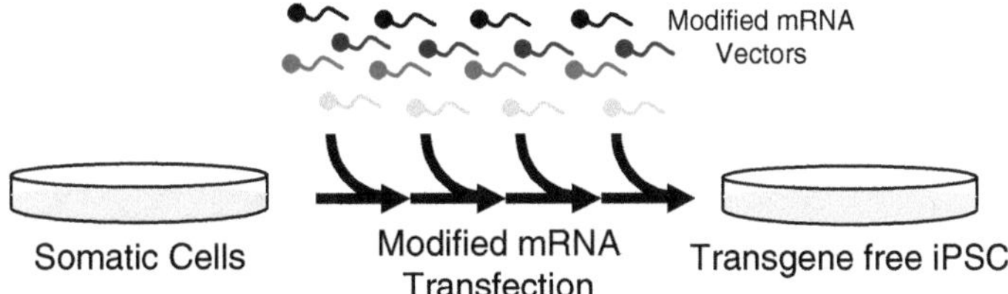

Fig. 6 Modified mRNA. Similar to transgene expression following transfection of recombinant DNA, transgenes can be expressed following the transfection of in vitro transcribed synthetic mRNA. The short half-life of mRNA once again necessitates serial transfections; however, the iPSCs produced by this protocol are transgene free and do not require extensive screening to ensure no exogenous DNA has become integrated

the dangers associated with DNA integration [25]. This method uses in vitro transcription (IVT) reactions utilizing PCR amplified templates encoding the four Yamanaka reprogramming factors (Oct4, Sox2, Klf4, and cMyc) with or without the addition of Lin28. To decrease the immunogenicity of transfected ssRNA (mediated by RIG-I, PKR, TLR7, and TLR8) the IVT reactions were carried out in the presence of unmodified and modified ribonucleotides (5-methylcytidine and pseudouridine) and an anti-reverse di-guanosine cap analog. The IVT reaction products were then DNAse treated to remove template DNA and phosphatase treated to remove 5′ triphosphates from residual uncapped synthetic mRNA. The protein expression from these highly stable synthetically produced mRNA's reached its peak within 12 h of transfection and significantly decreased thereafter. Therefore, for the continuous high protein expression required for reprogramming daily mRNA transfection is required for up to 18 days (Fig. 6) [25].

The efficiency of synthetic mRNA reprogramming ranges from 0.6 to 4.4 %. The highest efficiency of reprogramming was obtained using transfection of all five factors (OSKML) in hypoxic (5 % O_2) culture conditions. mRNA reprogramming efficiency was compared to traditional retroviral reprogramming and was found to be increased 0.04 % versus 1.4 %, respectively, and proceed with faster kinetics (ES like colonies appeared in 13–15 days vs. 25–29 days respectively) [25].

Reprogramming of human cells with modified synthetic mRNAs is an efficient process that results in transgene free iPSCs, but it is not without its pitfalls. Specifically, mRNA mediated reprogramming is a complex process requiring multiple steps (mRNA production, purification, transfection, etc.), numerous quality control measures and daily mRNA transfections to maintain high protein expression. Additionally, mRNA reprogramming may not benefit from the addition of small molecules as much as traditional viral reprogramming, as demonstrated by limited increases in efficiency when valproic acid (VPA) was included in reprogramming experiments [25].

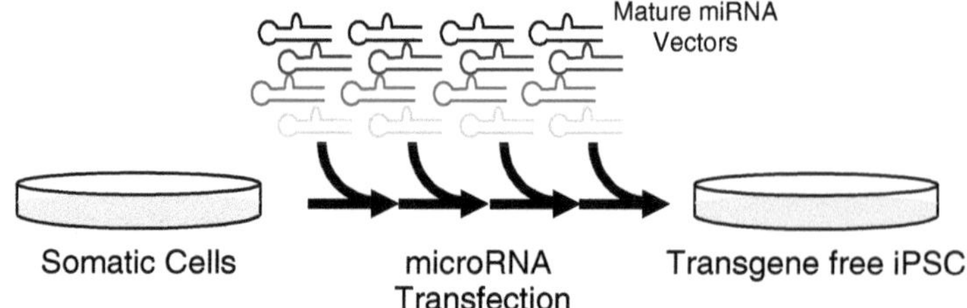

Fig. 7 Micro RNAs. Similar to transfected mRNAs, transfected miRNAs are also capable of reprogramming somatic cells. The protocol described by Miyoshi et al. requires four serial transfections of miRNAs to reprogram somatic cells. Again similar to mRNA reprogrammed cells miRNA reprogrammed iPSCs are transgene free and do not require extensive screening to ensure no exogenous DNA has become integrated

4.2 MicroRNAs

For years it has been known that microRNAs (miRNA), small non-coding RNAs that are thought to fine tune protein expression via binding with the 3′UTR of mRNA, play a role in maintaining pluripotency. It has also been shown that incorporation of miRNAs into traditional reprogramming strategies results in an increased reprogramming efficiency. The true impact that miRNAs can have on reprogramming was demonstrated in 2011 when Miyoshi et al. [26] generated iPSCs from mouse and human somatic cells using only miRNAs. Although previous studies had successfully reprogrammed mouse and human somatic cells using miRNA these miRNAs were expressed from integrated viral vectors [27]. miRNA mediated reprogramming was accomplished via transfection of a cocktail of three mature miRNAs: miR-200c, miR-302s, and miR-369s. This miRNA cocktail was used to successfully reprogram mouse adipose stromal cells (mASC), mouse embryonic fibroblasts (MEF), human adipose stromal cells (hASC) and human dermal fibroblasts (hDF). This reprogramming protocol calls for four serial transfections of the miRNA reprogramming cocktail every 48 h. After the eighth day, transfection cycle cells are transferred to ESC culture conditions and ESC like colonies begin to appear by 20 days after the first transfection (Fig. 7). The efficiency of miRNA mediated reprogramming of mASC is similar to the original retroviral mediated reprogramming of MEFs at 0.01 % and, as one might expect, reprogramming of human cells is less efficient with a frequency of two iPSC colonies per 1×10^5 starting cells (0.002 %) [26].

Similar to other non-DNA mediated reprogramming strategies the miRNA technique requires no screening to exclude exogenous DNA sequences being incorporated into iPSC clones. Additionally, the reprogramming efficiency is not substantially decreased as compared to other non-integrative methods. Although, multiple miRNAs and multiple transfections are required to reprogram cells, Miyoshi et al. was able to efficiently induce pluripotency without the use of small molecules.

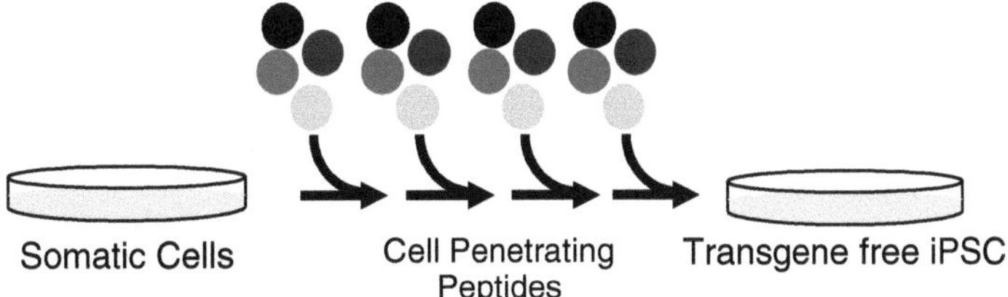

Fig. 8 Recombinant proteins. Reprogramming factors tagged with a poly-arginine cell penetrating peptide are capable of entering the nuclei of somatic cells and inducing pluripotency. Reprogramming using these recombinant cell penetrating peptides requires multiple protein treatments to dedifferentiate cells. By using proteins and not nucleic acids the iPSCs produced by this method do not need to be extensively screened for exogenous transgenes

4.3 Recombinant Proteins

Similar to transfection of modified synthetic mRNAs, the introduction of recombinant reprogramming proteins themselves is sufficient to induce pluripotency.

In an attempt to develop transgene free iPSCs for their potential clinical use Kim et al. and Zhou et al. both developed systems where recombinant reprogramming proteins (OSKM) tagged with a polyarginine cell penetrating peptide (CPP) could be used, with or without the addition of valproic acid (VPA), to reprogram murine embryonic fibroblasts (MEF) and human newborn fibroblasts (HNF) respectively (Fig. 8) [28, 29]. These two protocols differ in that one method uses bacterial protein expression followed by isolation, solubilization, refolding and purification steps and the other uses a total cell extract isolated from eukaryotic cells expressing the CPP tagged reprogramming factors.

In Kim et al.'s approach, MEFs were transduced with bacterially expressed recombinant proteins in the presence of valproic acid (VPA) four times over a 9 day period. 21–26 days after protein transduction ESC like colonies were picked and further expanded. Similar to other methods of reprogramming, this protocol is also relatively inefficient. Even in the presence of valproic acid, a HDAC inhibitor shown to increase reprogramming efficiency ~100-fold, only three iPSC clones were isolated from 5×10^4 starting cells, an efficiency of 0.006 % [28].

Zhou et al.'s approach uses retrovirally transduced HEK293 cells to express the CPP tagged reprogramming factors. Total cell extracts isolated from reprogramming factor expressing HEK293 cells were then used to reprogram HNFs in the absence of small molecules. In order to avoid cytotoxicity associated with prolonged exposure to the total cell extract, the authors used multiple cycles of 16 h protein treatment followed by 6 days of culture in ESC media. A minimum of six cycles was required before alkaline phosphatase positive cells could be detected, a total of 42 days of protein transduction. 14 days after the end of the last protein transduction cycle ESC like colonies were picked. Using this protocol the authors were able to isolate five iPSC clones from 5×10^5 starting cells, an efficiency of 0.001 %. Whole protein extracts were used and titration of reprogramming factors was not performed, so this protocol must still be optimized. Additionally, reprogramming was accomplished in the absence of any chemical treatments so treatment with small molecules may increase reprogramming efficiency as seen with other reprogramming protocols [29].

Recombinant protein mediated reprogramming does not require extensive screening to rule out the introduction of exogenous genetic material resulting from integration events or inefficient transgene removal. However, both protocols require multiple protein transductions (4–6) over an extended period of time to achieve reprogramming, and neither protocol is as efficient as the modified mRNA approach.

5 Small Molecules

5.1 Introduction

Cellular reprogramming is an immensely time and cost intensive process. Multiple methods can be used to increase reprogramming efficiency including the use of a starting cell type that endogenously expresses one or more of the traditional reprogramming factors. Isolation of these cell types is often more difficult than dermal fibroblast isolation, would be unethical to isolate from humans, require special culturing conditions or potentially increase the duration of an already time consuming process. For these reasons, multiple groups have investigated the potential of the myriad of small molecules available and found small molecules capable of increasing reprogramming efficiency, replacing traditional reprogramming factors and able to reprogram specific cell types without the need of exogenous reprogramming proteins or nucleic acids.

Poor reprogramming efficiencies can be overcome by including a variety of small molecules. However, the inclusion of small molecules is not without cost. It has been proposed that the use of small molecules, many of which are oncogenic, may induce mutations and thus promote neoplastic development following the differentiation of iPSCs. Similar to the use of non-integrating DNA protocols, where genome wide screening is needed to ensure exogenous DNA sequences have not been introduced, genome wide sequencing may be required before iPSCs reprogrammed in the presence of small molecules before they can be developed for clinical use.

5.2 Histone Deacetylase Inhibitors

One of the most commonly used small molecules is valproic acid (VPA), a histone deacetylase (HDAC) inhibitor [30]. VPA has been used to increase reprogramming efficiencies up to ~100-fold [30]. VPA is also capable of replacing both Klf4 and cMyc, albeit at lower efficiencies than OSKM+VPA [31]. However, it has recently been shown that another HDAC, sodium butyrate (NaB), is more efficient at increasing the reprogramming efficiency and the combination of NaB with PS48 (an activator of 3′-phosphoinositide dependent kinase that is capable of increasing reprogramming efficiency by itself) is able to further increase the efficiency over NaB mediated reprogramming [32].

5.3 Two Inhibitor (2i) Cocktail

The small molecules PD0325901 (a mitogen activated protein kinase/ERK kinase inhibitor) and CHIR99021 (a glycogen synthase kinase inhibitor) are commonly used in combination to promote ground state pluripotency and the growth of true iPSC colonies. It has also been shown that this small molecule combination is able to maintain the self-renewal properties of both ESCs and iPSCs in the absence of leukemia inhibitory factor (LIF) and serum/Bone Morphological Protein (BMP) [33]. In combination with or independent of CHIR99021, PD0325901 can facilitate the conversion of cells stuck in a pre-iPSC state into a true iPSC state, and independent of PD0325901, CHIR99021 has been demonstrated to replace Sox2 in OKM reprogramming experiments [33–35].

5.4 Kenpaullone

A large chemical screen looking at the ability of small molecules to reprogram cells in the presence of exogenous OSM found that kenpaullone, an inhibitor of cyclin dependent kinases 1, 2, 5 and GSK3, was able to replace Klf4 [36]. Similar to CHIR99021, kenpaullone was also able to increase the efficiency of four factor reprogramming. However, when CHIR99021 alone or in combination with purvalanol A (a promiscuous cyclin dependent kinase inhibitor) was used in the same OSM screen, no iPSC colonies were found, suggesting that the reprogramming activity of kenpaullone is mediated through a non-GSK pathway [36].

5.5 TGF-β Inhibitors

Multiple groups have shown that the inhibition of the transforming growth factor β (TGF- β) pathway (e.g., with the small molecules A-83-01 or RepSox) can increase reprogramming efficiency and replace Sox2 and/or cMyc in reprogramming experiments [37, 38]. Interestingly inhibitors of the Pan-Src kinase family (e.g., iPYrazine, Dasatanib, or PP1) can also replace Sox2 in reprogramming experiments [39]. Furthermore, AMI-5 mediated protein arginine methyltransferase inhibition coupled with AM-83-01 mediated TGB-β inhibition is capable of reprogramming MEFs transduced with Oct4 only [40].

5.6 Histone Methyltransferase Inhibition

Similar to other small molecules BIX01294, a G9a histone methyltransferase inhibitor, has been shown to replace Sox2 in reprogramming experiments [35, 41]. However, unlike other small molecules, BIX01294 can replace Oct4 in reprogramming experiments, but this replacement of Oct4 was demonstrated in mouse neural progenitor cells, a cell type with increased endogenous reprogramming factor expression. BIX01294 is also capable of increasing reprogramming efficiency and BIX01294 mediated increases in reprogramming efficiency are further increased with inclusion of BayK, an L-type calcium channel agonist [41].

5.7 ROCK Inhibition

The ROCK kinase inhibitor Y27632 is commonly used to increase survival of trypsinized pluripotent cells in reprogramming protocols and passage of ESC and iPSCs [42]. Lin et al. [43] demonstrated that thiazovivin is also able to increase survival post trypsinization, and that a combination of thiazovivin, PD0325901 and the ALK5 inhibitor SB431542 is capable of increasing reprogramming efficiency by more than 200-fold over OSKM transduction alone.

5.8 DNA Methyl Transferase Inhibitors

Similar to HDAC inhibitors, inhibitors of DNA methyl transferase such as AZA and RG108 are capable of increasing reprogramming efficiency up to ~10-fold [44]. Even more interesting is that single treatment of myoblasts, cells that endogenously express Sox2, Klf4, and cMyc, with RG108 produced iPSCs within 3 weeks [45]. These single small molecule reprogrammed cells are capable of contributing to all three germ layers and were differentiated into cardiomyocytes capable of improving cardiac function following infarction.

6 Conclusions

The field of cellular reprogramming is fast paced and constantly evolving. New reprogramming protocols and methods to increase both efficiency and quality are constantly being developed. Non-integrative, non-DNA techniques are the way of the future, but they are hampered by complex manipulations, special reagents, and low reprogramming efficiencies. Optimization of these the protocols is needed before they can be widely accepted.

Cellular reprogramming has come a long way since the SCNT experiments first done in the 1960s, but a significant amount of work is still needed to develop a technique that can efficiently produce safe patient-specific iPSCs for potential use in a clinical setting.

Acknowledgments

This work would not have been possible without the support of a VA Merit Award grant No. 1I01BX001125-02 and grant NIH/NHLBI 2 R01 HL073015-07 to N.Z. We would also like to thank Gohar Manzar for her helpful suggestions in preparing this manuscript.

References

1. Martin GR (1981) Isolation of a pluripotent cell line from early mouse embryos cultured in medium conditioned by teratocarcinoma stem cells. Proc Natl Acad Sci USA 78(12): 7634–7638
2. Thomson JA et al (1998) Embryonic stem cell lines derived from human blastocysts. Science 282(5391):1145–1147
3. Briggs R, King TJ (1952) Transplantation of living nuclei from blastula cells into enucleated frogs' eggs. Proc Natl Acad Sci USA 38(5):455–463
4. Gurdon JB, Wilmut I (2011) Nuclear transfer to eggs and oocytes. Cold Spring Harb Perspect Biol 3(6):a002659
5. Gurdon JB, Byrne JA, Simonsson S (2003) Nuclear reprogramming and stem cell creation. Proc Natl Acad Sci USA 100(Suppl 1):11819–11822
6. Takahashi K, Yamanaka S (2006) Induction of pluripotent stem cells from mouse embryonic and adult fibroblast cultures by defined factors. Cell 126(4):663–676
7. Takahashi K et al (2007) Induction of pluripotent stem cells from adult human fibroblasts by defined factors. Cell 131(5):861–872
8. Yu J et al (2007) Induced pluripotent stem cell lines derived from human somatic cells. Science 318(5858):1917–1920
9. Okita K, Ichisaka T, Yamanaka S (2007) Generation of germline-competent induced pluripotent stem cells. Nature 448(7151): 313–317
10. Aoi T et al (2008) Generation of pluripotent stem cells from adult mouse liver and stomach cells. Science 321(5889):699–702
11. Wang W et al (2011) Rapid and efficient reprogramming of somatic cells to induced pluripotent stem cells by retinoic acid receptor gamma and liver receptor homolog 1. Proc Natl Acad Sci USA 108(45):18283–18288
12. Maekawa M et al (2011) Direct reprogramming of somatic cells is promoted by maternal transcription factor glis1. Nature 474(7350): 225–229
13. Mali P et al (2008) Improved efficiency and pace of generating induced pluripotent stem cells from human adult and fetal fibroblasts. Stem Cells 26(8):1998–2005
14. Carey BW et al (2009) Reprogramming of murine and human somatic cells using a single polycistronic vector. Proc Natl Acad Sci USA 106(1):157–162
15. Kaji K et al (2009) Virus-free induction of pluripotency and subsequent excision of reprogramming factors. Nature 458(7239):771–775
16. Woltjen K et al (2009) piggyBac transposition reprograms fibroblasts to induced pluripotent stem cells. Nature 458(7239):766–770
17. Stadtfeld M et al (2008) Induced pluripotent stem cells generated without viral integration. Science 322(5903):945–949
18. Zhou W, Freed CR (2009) Adenoviral gene delivery can reprogram human fibroblasts to induced pluripotent stem cells. Stem Cells 27(11):2667–2674
19. Okita K et al (2008) Generation of mouse induced pluripotent stem cells without viral vectors. Science 322(5903):949–953
20. Si-Tayeb K et al (2010) Generation of human induced pluripotent stem cells by simple transient transfection of plasmid DNA encoding reprogramming factors. BMC Dev Biol 10:81
21. Yu J et al (2009) Human induced pluripotent stem cells free of vector and transgene sequences. Science 324(5928):797–801
22. Conese M, Auriche C, Ascenzioni F (2004) Gene therapy progress and prospects: episomally maintained self-replicating systems. Gene Ther 11(24):1735–1741
23. Chou BK et al (2011) Efficient human iPS cell derivation by a non-integrating plasmid from blood cells with unique epigenetic and gene expression signatures. Cell Res 21(3): 518–529
24. Jia F et al (2010) A nonviral minicircle vector for deriving human iPS cells. Nat Methods 7(3):197–199
25. Warren L et al (2010) Highly efficient reprogramming to pluripotency and directed differentiation of human cells with synthetic modified mRNA. Cell Stem Cell 7(5): 618–630
26. Miyoshi N et al (2011) Reprogramming of mouse and human cells to pluripotency using mature microRNAs. Cell Stem Cell 8(6): 633–638
27. Anokye-Danso F et al (2011) Highly efficient miRNA-mediated reprogramming of mouse and human somatic cells to pluripotency. Cell Stem Cell 8(4):376–388
28. Kim D et al (2009) Generation of human induced pluripotent stem cells by direct delivery of reprogramming proteins. Cell Stem Cell 4(6):472–476
29. Zhou H et al (2009) Generation of induced pluripotent stem cells using recombinant proteins. Cell Stem Cell 4(5):381–384
30. Huangfu D et al (2008) Induction of pluripotent stem cells by defined factors is greatly improved by small-molecule compounds. Nat Biotechnol 26(7):795–797

31. Huangfu D et al (2008) Induction of pluripotent stem cells from primary human fibroblasts with only Oct4 and Sox2. Nat Biotechnol 26(11):1269–1275
32. Zhu S et al (2010) Reprogramming of human primary somatic cells by OCT4 and chemical compounds. Cell Stem Cell 7(6):651–655
33. Silva J et al (2008) Promotion of reprogramming to ground state pluripotency by signal inhibition. PLoS Biol 6(10):e253
34. Li W et al (2009) Generation of human-induced pluripotent stem cells in the absence of exogenous Sox2. Stem Cells 27(12): 2992–3000
35. Shi Y et al (2008) A combined chemical and genetic approach for the generation of induced pluripotent stem cells. Cell Stem Cell 2(6): 525–528
36. Lyssiotis CA et al (2009) Reprogramming of murine fibroblasts to induced pluripotent stem cells with chemical complementation of Klf4. Proc Natl Acad Sci USA 106(22): 8912–8917
37. Ichida JK et al (2009) A small-molecule inhibitor of tgf-Beta signaling replaces sox2 in reprogramming by inducing nanog. Cell Stem Cell 5(5):491–503
38. Li W et al (2009) Generation of rat and human induced pluripotent stem cells by combining genetic reprogramming and chemical inhibitors. Cell Stem Cell 4(1):16–19
39. Staerk J et al (2011) Pan-Src family kinase inhibitors replace Sox2 during the direct reprogramming of somatic cells. Angew Chem 50(25):5734–5736
40. Yuan X et al (2011) Combined chemical treatment enables Oct4-induced reprogramming from mouse embryonic fibroblasts. Stem Cells 21(6):884–894
41. Shi Y et al (2008) Induction of pluripotent stem cells from mouse embryonic fibroblasts by Oct4 and Klf4 with small-molecule compounds. Cell Stem Cell 3(5):568–574
42. Watanabe K et al (2007) A ROCK inhibitor permits survival of dissociated human embryonic stem cells. Nat Biotechnol 25(6): 681–686
43. Lin T et al (2009) A chemical platform for improved induction of human iPSCs. Nat Methods 6(11):805–808
44. Mikkelsen TS et al (2008) Dissecting direct reprogramming through integrative genomic analysis. Nature 454(7200):49–55
45. Pasha Z, Haider H, Ashraf M (2011) Efficient non-viral reprogramming of myoblasts to stemness with a single small molecule to generate cardiac progenitor cells. PLoS One 6(8): e23667

Chapter 7

Differentiation and Lineage Commitment of Murine Embryonic Stem Cells into Insulin Producing Cells

Sudhanshu P. Raikwar and Nicholas Zavazava

Abstract

Pluripotent embryonic stem (ES) cells and induced pluripotent stem (iPS) cells recently developed in our laboratory can be used to generate the much needed insulin producing cells (IPCs) for the treatment of type 1 diabetes. However, currently available differentiation protocols generate IPCs at a very low frequency. More importantly, it is difficult to purify the IPCs from the mixed cell population due to the lack of well characterized pancreatic beta cell-specific cell surface markers. Subsequently, multiple studies have been published with limited success. A major cause for these poor results is an inadequate Pdx1 expression in the embryoid body (EB) or definitive endoderm (DE)-derived precursors. Here we investigated whether ectopic expression of pancreatic and duodenal homeobox 1 (Pdx1), an essential pancreatic transcription factor, in mouse ES cells leads to enhanced differentiation into IPCs. Here we describe a new approach for the generation of glucose responsive IPCs using ES cells ectopically expressing pancreatic and duodenal homeobox 1 (Pdx1) and paired box gene 4 (PAX4).

Key words Activin A, Baculovirus, Definitive endoderm, Embryonic stem cells, Diabetes, Differentiation, Insulin producing cells, Lentiviral vectors, PAX4, Pdx1

1 Introduction

Type 1 diabetes is an autoimmune disease that can be medically treated or surgically corrected by either whole pancreas or islet transplantation [1–4]. However, chronic shortage of organ donors, lifelong immunosuppressive therapy and chronic graft rejection currently limit the therapeutic potential of islet transplantation. Ultimately chronic graft rejection leads to insulin dependence and the development of serious diabetic complications [5–10]. With the incidence of diabetes increasing worldwide at an alarming rate, there is an urgent and compelling need to develop novel forms of treatment for diabetes. In this regard, pluripotent embryonic stem (ES) cells and the recently developed induced pluripotent stem (iPS) cells offer a novel approach for the development of stem cell based therapies [11–14]. However, due to complex molecular

Nicholas Zavazava (ed.), *Embryonic Stem Cell Immunobiology: Methods and Protocols*, Methods in Molecular Biology, vol. 1029, DOI 10.1007/978-1-62703-478-4_7, © Springer Science+Business Media New York 2013

mechanisms underlying pancreatic β cell development that are not yet very well understood, the generation of functional glucose responsive IPCs from the mouse and human ES and iPS cells has not been achieved satisfactorily.

Prior work has shown that the mouse ES cells poorly differentiate into IPCs [15–21] and consequently fail to correct hyperglycemia in diabetic mice. Hence, there is an urgent need to develop robust protocols for the differentiation of mouse ES cells into IPCs. During embryonic development, a wide variety of transcription factors including pancreatic and duodenal homeobox 1 (Pdx1) and paired box gene 4 (PAX4) are involved in pancreatic β cell differentiation. Pdx1 is involved in the normal development of the pancreas [22–25] during embryogenesis, β-cell differentiation [23, 26–28] and is essential for the maintenance of β-cell function in the adult [23, 29–31]. Pdx1 mutation in human and mice causes failure of pancreas development leading to diabetes [22, 32]. PAX4 expression is restricted to the β and δ cell lineages within the developing pancreas and mice lacking PAX4 fail to develop any β cells and are diabetic [22, 33–36]. Here, we hypothesized that lineage commitment of ES cells by pancreatic transcription factors enhances the generation of IPCs thereby providing an unlimited source of cells for the treatment of diabetes. To test our hypothesis, a novel approach for in vitro ES cell differentiation into IPCs was developed. We transduced ES cell-derived Nestin$^+$ cells with feline immunodeficiency virus (FIV) based lentiviral vectors expressing either Pdx1 or PAX4 transcription factors which are critical for the development of pancreatic islets. The rationale underlying this approach is that expression of pancreatic β-cell-specific transcription factors will maximize the lineage commitment and differentiation of ES and iPS cells into IPCs [16, 24, 29, 35, 37–42]. Ectopic expression of Pdx1 in the Nestin$^+$ cells led to the robust generation of IPCs while PAX4 had only minimal effect on the development of IPCs. However, the major limitation of this approach is that it requires lentiviral transduction of Nestin$^+$ cells. To facilitate greater efficiency, we have now generated a double transgenic ES cell line R1Pdx1AcGFP/RIP-Luc to stably express a Pdx1AcGFP fusion protein [40].

2 Materials

Mouse ES cells (ATCC, Manassas, VA), γ irradiated primary mouse embryonic fibroblasts (Millipore, Billerica, MA), ES cell medium [(High glucose DMEM medium (Invitrogen, Carlsbad, CA), 1,000 U/ml ESGRO LIF (Millipore, Billerica, MA), 1 % nonessential amino acids (Invitrogen, Carlsbad, CA), 1 % nucleosides (Millipore, Billerica, MA), 1 % β-mercaptoethanol, 1 % l-Glutamine, 1 % penicillin–streptomycin (Invitrogen, Carlsbad, CA),

15–20 % ES qualified FBS (Hyclone, Ogden, UT)], porcine gelatin (Sigma, St. Louis, MO), collagen type IV (BD Bioscience, Franklin Lakes, NJ), α-Monothioglycerol (Sigma, St. Louis, MO), HEPES buffer (Invitrogen, Carlsbad, CA), glutamine (Invitrogen, Carlsbad, CA), sodium pyruvate (Invitrogen, Carlsbad, CA), insulin transferrin selenium (ITS) mix (Invitrogen, Carlsbad, CA), N2 (Invitrogen, Carlsbad, CA), B27 (Invitrogen, Carlsbad, CA), bFGF (Invitrogen, Carlsbad, CA), EGF (Invitrogen, Carlsbad, CA), KGF (Invitrogen, Carlsbad, CA), KRBH buffer, nicotinamide (Sigma, St. Louis, MO), ultrasensitive mouse insulin ELISA kit (Mercodia, Winston salem, NC), tolbutamide (Sigma, St. Louis, MO), nifedipine (Sigma, St. Louis, MO), primary and secondary antibodies (Santa Cruz Biotechnology, Santa Cruz, CA), immunohistochemical and immunostaining reagents (Dako North America, Carpinteria, CA), flow cytometry permeabilization kit (BD Biosciences, Franklin Lakes, NJ), lentiviral vectors (University of Iowa Gene Transfer Vector Core, Iowa City, IA and Genecopoeia, Rockville, MD), Bac-to-Bac baculovirus expression vector kit (Invitrogen, Carlsbad, CA), pCMV6-AC expression vector (SC319183, OriGene Technologies, Rockville, MD), 12 % precast SDS PAGE gels (Bio-Rad, Hercules, CA), Laemmli buffer for cell lysis (Bio-Rad, Hercules, CA), Prestained Precision Plus protein standard (Bio-Rad, Hercules, CA), His GraviTrap columns (GE Healthcare Life Sciences, Piscataway, NJ), PD10 columns (GE Healthcare Life Sciences, Piscataway, NJ), Amicon Ultra 4-centrifugal filter units (Millipore, Billerica, MA), Millex syringe filters (Millipore, Billerica, MA), ultralow attachment tissue culture plates (Corning, Lowell, MA), six well tissue culture plates (Corning, Lowell, MA), 10 and 15 cm^2 tissue culture plates (Corning, Lowell, MA), T25 and T75 cm^2 tissue culture flasks (Corning, Lowell, MA), 5, 10, and 15 ml pipets (Corning, Lowell, MA), cell scrapers (Corning, Lowell, MA)].

3 Methods

3.1 Generation of FIV Based Lentiviral Vectors Expressing Pdx1 and Pax4

The generation of vesicular stomatitis virus (VSV) envelope pseudotyped FIV based lentiviral vectors expressing Pdx1 was achieved using the following strategy.

1. For generating the FIV based lentiviral vectors, first the Pdx1-AcGFP coding sequences were excised from pCMV-Pdx1-AcGFP as EcoRI-HpaI fragment, treated with Klenow polymerase to generate blunt ends and subcloned into lentiviral transfer vector pVETLCSKh10 (University of Iowa Gene Transfer Vector Core, Iowa City, IA) plasmid at the EcoRV site to generate pVETLC-Pdx1-AcGFP (*see* **Note 1**).

2. The lentivirus FIV-Pdx1-AcGFP was generated by triple transfection of 293T cells with vector construct pVETLC-Pdx1-AcGFP, envelope plasmid pCMV-VSV-G and the FIV packaging plasmid pCFIVΔorf2Δvif (University of Iowa Gene Transfer Vector Core, Iowa City, IA) (*see* **Note 2**).
3. The other lentiviral vectors FIV-CMV-eGFP and FIV-PAX4-FLAG were generated using the same strategy but employing either pVETLC-eGFP University of Iowa Gene Transfer Vector Core, Iowa City IA) or pReceiver-Lv03 (EX-P0096-Lv03, GeneCopoeia, Rockville, MD) containing an inframe fusion between human PAX4 and FLAG tag at the C-terminus of PAX4 (NM_006193) constructs, respectively (*see* **Note 3**).
4. The vector containing medium was collected at 24, 36, and 72 h, by replacing this medium with fresh DMEM at 24 and 36 h time points. At each collection the medium was filtered through a 0.45 μm bottle top filter and stored at 4 °C.
5. The viral particles were concentrated by centrifuging the collected medium at 4 °C for 16 h at 5911 ×*g*. The viral pellet was carefully resuspended in PBS containing lactose (40 mg/l).

3.2 Generation of ES Cell-Derived IPCs

The undifferentiated R1 ES cells were subjected to differentiation using a multistep modified Lumelsky's protocol [15] (*see* **Note 4**).

1. Cultivate mouse R1 ES cells in High glucose DMEM supplemented with LIF (1,000 U/ml) on a semiconfluent monolayer of γ irradiated primary mouse embryonic fibroblast feeder cells (*see* **Note 5**). ES cells must be maintained in fresh medium every 2–3 days. ES cells must be passaged every 2–3 days since frequent passaging removes the differentiated cells (*see* **Notes 6–8**).
2. The undifferentiated ES cells were trypsinized and 1×10^7 cells were directly plated on to ultra-low attachment tissue culture dishes in the presence of freshly prepared (45 μl/50 ml) 1:10 α-Monothioglycerol (Sigma chemical Company, St. Louis, MO) to promote EB formation for 4 days. Alternatively, EBs can be generated by hanging drop method by plating 25 μl medium containing 400 ES cells/drop multiple times on to the inner surface of the lid of a 15 cm^2 tissue culture dish and keeping it inverted in the tissue culture incubator for 2 days. Once the EBs form, they can be isolated from the hanging drops and kept in ultra-low attachment tissue culture dishes for 2 days.
3. The EBs were washed in PBS and plated on to gelatin coated plates at low density in DMEM supplemented with 10 % FBS for 6 days. Following attachment of the EBs on to the plastic surface, they progressively become flattened and spread out within 24 h.

4. Lentiviral transduction of EB-derived Nestin⁺ cells was performed at a multiplicity of infection (MOI) of ten transducing units/cell using either FIV-CMV-eGFP or FIV-Pdx1-AcGFP or FIV-Pax4-FLAG. The lentiviral transduced Nestin⁺ cells were further differentiated to generate IPCs (*see* **Note 9**).
5. The Nestin⁺ cells were grown in DMEM/F12 (1:1) medium supplemented with 25 ng/ml bFGF (R&D System Inc. Minneapolis, MN), N2 and B27 (Invitrogen, Carlsbad, CA) supplements and cultured for 9 days to generate endocrine precursors.
6. The endocrine precursors were further propagated for 6 days in low glucose DMEM supplemented with N2, B27, 10 ng/ml bFGF, EGF, and KGF, and 10 mM Nicotinamide to enrich IPCs.
7. The resulting IPCs were cultured in the presence of DMEM supplemented with N2, B27, and 10 mM nicotinamide for 13 days.
8. The IPCs generated with and without lentiviral transduction were characterized by immunostaining and tested for their ability to secrete insulin by ELISA (Figs. 1, 2, and 3). The undifferentiated ES cells, EBs as well as ES cell-derived IPCs expressed very low levels of MHC class I and class II molecules (Fig. 4) thereby making them poorly immunogenic and susceptible to NK cell-mediated killing.

3.3 Generation of Recombinant Baculovirus Expressing Human Activin A

Baculovirus expression vectors have increasingly become one of the most powerful and widely used systems for the production of recombinant proteins [43]. We have achieved a high level expression of recombinant human activin A using the commercially available Bac-to-Bac baculovirus expression system (Invitrogen, Carlsbad, CA). Generation of recombinant baculovirus is a multistep process. The following scheme was used to generate the recombinant baculovirus.

1. The 1,281 bp human activin A coding sequence consists of a secretory signal peptide followed by the activin A coding sequence.
2. The cDNA encoding human activin a was modified by PCR amplification using pCMV6-AC expression vector (SC319183, OriGene Technologies, Rockville, MD) and the following PCR primers:

   Forward primer:

   5′-CGGAATTCATGCCCTTGCTTTGGCTGAGAGGATT TCTGTTGGC-3′.

   Reverse primer:

   5′-GCTCTAGACTAGTGATGGTGATGGTGATGTGAGC ACCCACACTCCTC-3′.

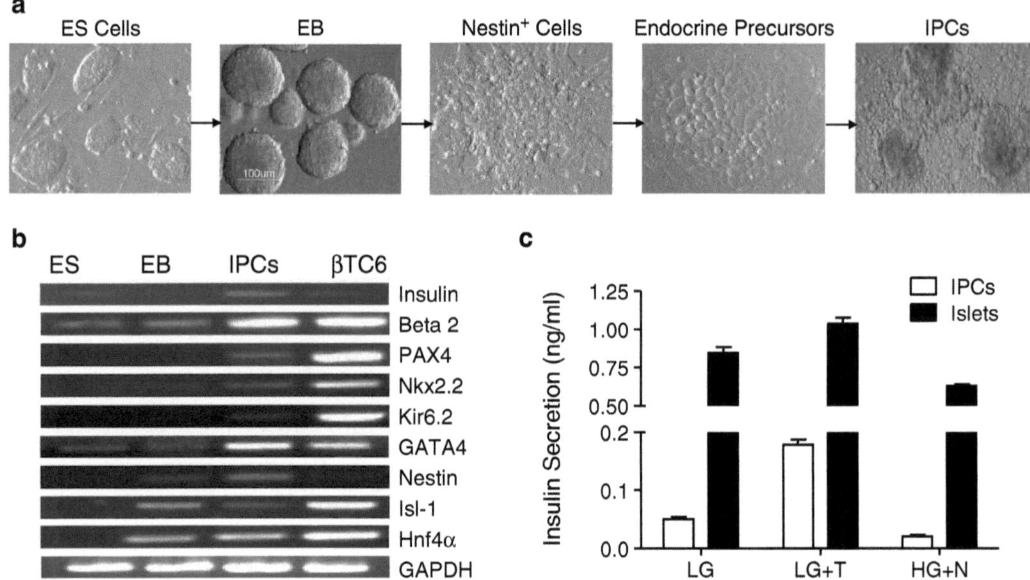

Fig. 1 Differentiation of ES cells into insulin producing cells (IPCs): (**a**) ES cells were subjected to differentiation using a multistep differentiation protocol and phase contrast pictures were taken at various time points. Embryoid body (EB) formation represents the first step during multistep differentiation events followed by Nestin+ intermediate stage, which ultimately leads to the development of IPCs via the development of an endocrine precursor stage. (**b**) Differential gene expression of ES cells undergoing differentiation into IPCs was examined by RT-PCR. Total RNA was isolated from the undifferentiated ES cells, EB, IPCs and βTC-6 cells and used for RT-PCR for several pancreatic β-cell-specific genes. The house keeping gene GAPDH was used as a positive control to normalize and validate the results. The RT-PCR results indicate gradual upregulation of pancreatic β-cell-specific genes selectively in the ES cell-derived IPCs during differentiation. Gene expression in IPCs is similar to that in the control βTC6 cells. (**c**) 1×10^6 ES cell-derived IPCs and ten mouse pancreatic islets were tested for their glucose responsive insulin secretion by ELISA in triplicates in the presence of low glucose (LG), LG + Tolbutamide (LG + T), and high glucose (HG) + Nifedipine (HG + N)

3. The modified human activin A cDNA includes an inframe fusion of 6× His Tag at the 3′ end for the purpose of purification of recombinant activin A by column chromatography (*see* **Note 10**).
4. Initially the activin A cDNA was subcloned as an EcoRI-XbaI fragment into the pFast/Bac Vector under the control of a baculovirus-specific strong polyhedrin (PH) promoter. The cloned activin A cDNA sequence was sequenced to rule out any errors in the coding sequence. The pFast/Bac vector encodes ampicillin as well as kanamycin antibiotic resistance markers for selection.
5. The pFastBac-Activin A expression vector was transformed into competent DH10Bac™ *E. coli*. DH10Bac™ cells that contain a baculovirus shuttle vector (bacmid) with a mini-attTn7 target site and a helper plasmid.

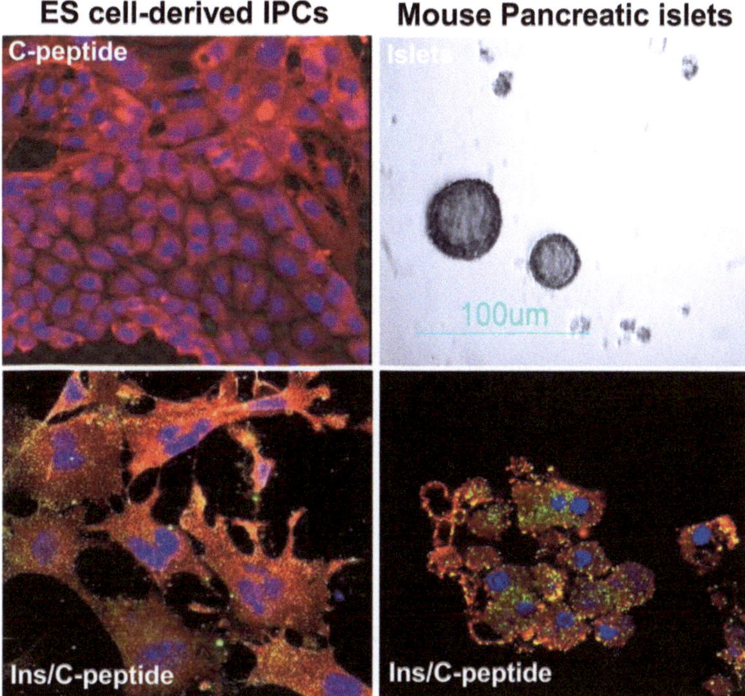

Fig. 2 Characterization of the ES cell-derived IPCs by immunostaining: the differentiated IPCs as well as isolated mouse pancreatic islets (positive control) were fixed in 2 % paraformaldehyde, permeabilized with 0.1 % Triton X100 and stained using primary anti-insulin and anti-C-peptide antibodies followed by staining using secondary conjugated antibodies. Analysis of the immunostaining pattern by multiphoton imaging revealed that ES cell-derived IPCs are positive for C-peptide (*red*) as well as insulin (*green*) which are specific markers of pancreatic β cells

6. Following transformation into DH10Bac™ cells, transposition occurs between the mini-Tn7 element on the pFastBac™ vector and the mini-attTn7 target site on the bacmid to generate a recombinant bacmid. This transposition reaction occurs in the presence of transposition proteins supplied by the helper plasmid.
7. The baculovirus shuttle vector (bacmid), bMON14272 (136 kb), present in DH10Bac™ *E. coli* contains, a low-copy number mini-F replicon, Kanamycin resistance marker and a segment of DNA encoding the LacZα peptide from a pUC-based cloning vector into which the attachment site for the bacterial transposon, Tn7 (mini-attTn7) has been inserted. Insertion of the mini-attTn7 attachment site does not disrupt the reading frame of the LacZα peptide.
8. The bacmid propagates in *E. coli* DH10Bac™ as a large plasmid that confers resistance to kanamycin and can complement

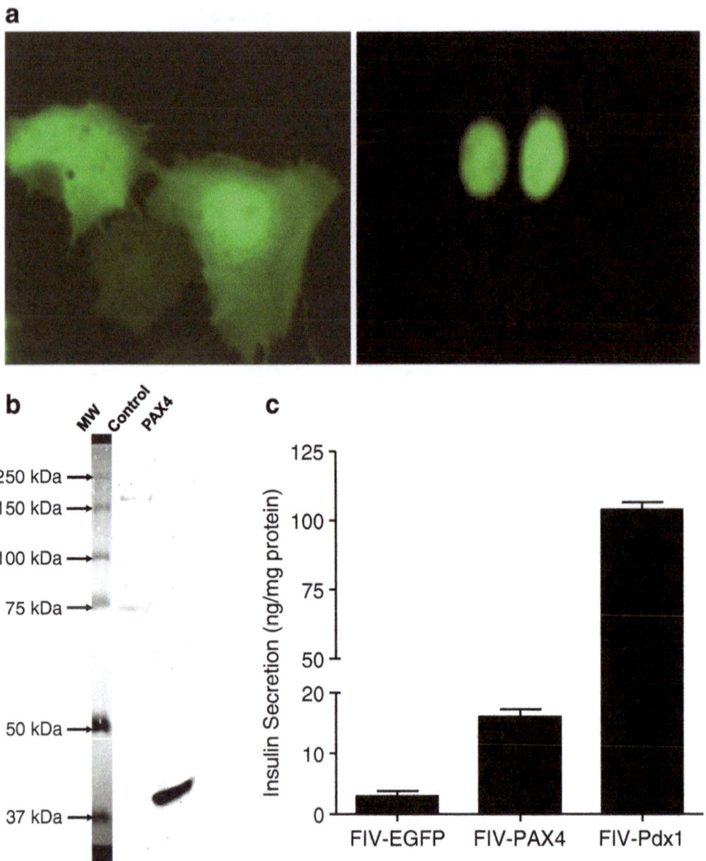

Fig. 3 FIV-based lentiviral vectors efficiently transduce Nestin⁺ cells and enhance the generation of IPCs: the ES cells were converted into EBs and allowed to differentiate into Nestin⁺ cells (**a**) which were transduced with lentiviral vectors FIV-EGFP, FIV-Pdx1, or (**b**) FIV-PAX4 expressing EGFP, Pdx1-AcGFP, and PAX4-FLAG, respectively. The transduced cells were differentiated into IPCs. (**c**) The insulin secretion in the supernatant was determined by ELISA and expressed as relative to the protein content of the cells. FIV-EGFP transduced cells expressed the least insulin, a moderate increase was observed following transduction with FIV-PAX4. The IPCs generated following FIV-Pdx1 transduction expressed the highest levels of insulin

a lacZ deletion present on the chromosome to form colonies that are blue (Lac^+) in the presence of a chromogenic substrate such as Bluo-gal or X-gal and the inducer, IPTG.

9. Recombinant bacmids (composite bacmids) are generated by transposing a mini-Tn7 element from a pFastBac™ donor plasmid to the mini-attTn7 attachment site on the bacmid. The Tn7 transposition functions are provided by a helper plasmid.
10. DH10Bac™ *E. coli* contains the helper plasmid, pMON7124 (13.2 kb), which encodes the transposase and confers resistance

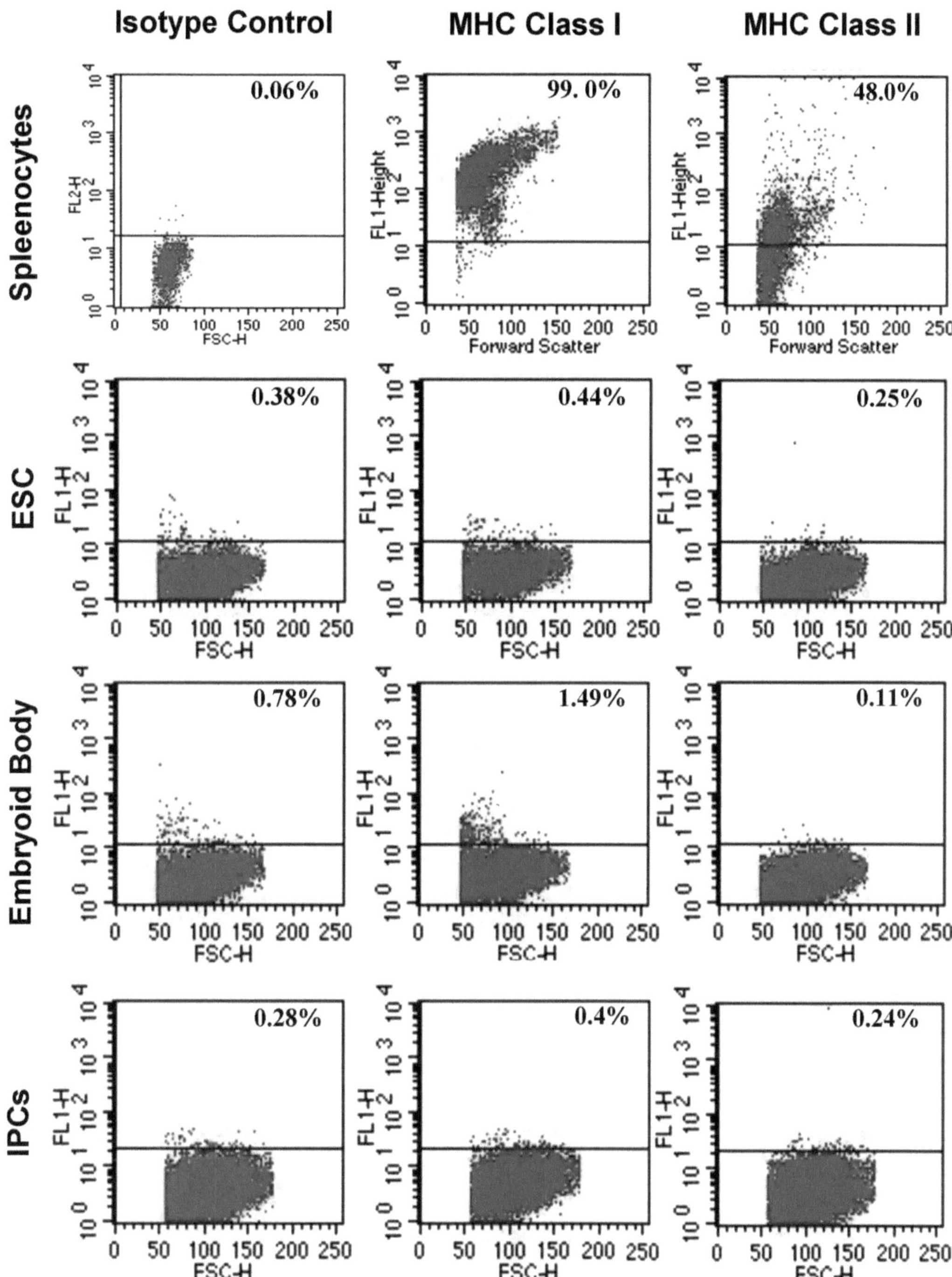

Fig. 4 ES cells and its derivatives are poorly immunogenic: mouse splenocytes (positive control), undifferentiated ES cells, embryoid body (EB) and ES cell-derived IPCs were stained using MHC class I and MHC class II antibodies. As compared to splenocytes, the undifferentiated ES cells, EB and ES cell-derived IPCs express low levels of MHC class I and MHC class II molecules

to tetracycline. The helper plasmid provides the Tn7 transposition function in trans.

11. Following homologous recombination in DH10Bac™ *E. coli* on LB plates containing Kanamycin (50 μg/ml), gentamycin (7 μg/ml), tetracycline (10 μg/ml), IPTG (40 μg/ml), and Bluo-Gal (300 μg/ml), the white colonies are selected.
12. Inoculate white colonies in 2 ml LB broth containing Kanamycin (50 μg/ml), gentamycin (7 μg/ml), tetracycline (10 μg/ml) and incubate at 37 °C in an orbital shaker at 250 rpm for a minimum of 16 h.
13. Gently isolate the recombinant bacmid DNA using the Concert High Purity Plasmid Miniprep system.
14. Plate SF9 insect cells on 6 well plates in SF-900II SFM medium and allow them to adhere for 30 min. Unlike mammalian cells, SF9 cells do not form a compact monolayer but quickly adhere to the cell culture plates/flasks.
15. Transfect the recombinant bacmid DNA using CellFectin reagent in the absence of antibiotics. Incubate the transfected cells for 5 h at 27 °C.
16. Remove the transfection mixture and add SF-900II SFM containing antibiotics and incubate at 27 °C for 72 h. Harvest the cell culture supernatant containing recombinant baculovirus.
17. Determine the titer of the recombinant baculovirus by performing plaque assay on SF9 cells. Calculate the titer as follows:

$$\text{Pfu/ml of original stock} = 1/\text{dilution factor} \times \text{number of plaques} \times 1/(\text{ml of inoculum/plate}).$$

18. Although it is essential to determine the titer of the recombinant baculovirus, plaque purification is not necessary with the site-specific transposition method.
19. Amplify the virus stock by infecting a suspension or monolayer culture in mid-exponential growth phase at an MOI of 0.01–0.1. Collect the supernatant containing concentrated recombinant baculovirus 48–72 h post-transduction.
20. Store the virus at 4 °C. The recombinant baculovirus is highly stable and maintains its integrity and infectious competency for several months to years at 4 °C.
21. For large scale production of recombinant human activin A, the SF9 cells grown in suspension culture at a density of 2×10^6–4×10^6 cells/ml are transduced with recombinant baculovirus at an MOI of 0.5–10.
22. The optimal expression of recombinant activin A takes place between 48 and 72 h post infection. Harvest the cell culture supernatant, centrifuge at 280 × *g* for 10 min to pellet the SF9

cells, and use the clarified supernatant to purify recombinant Activin A.

23. His-tagged recombinant activin A has a high selective affinity for Ni^{2+} and is easily purified using His GraviTrap flow purification (GE Healthcare) (*see* **Notes 11** and **12**).
24. Desalt the purified recombinant activin A using the PD10 (Amersham Biosciences) desalting columns (*see* **Note 13**).
25. Concentrate the purified protein using the Amicon Ultra 4-centrifugal filter units.
26. The purity of the recombinant activin A was confirmed by western blot analysis using anti-His antibody.

3.4 Generation of Definitive Endodermal (DE) Cells

The critical step wise progression that needs to be modeled in order to generate IPCs include the induction of definitive endoderm, the patterning and specification of endoderm to a pancreatic fate and the generation of mature endocrine cells. Multiple studies have demonstrated that induction of definitive endoderm is dependent on signaling through nodal pathway. We have optimized a highly reproducible and simple protocol for the generation of DE cells using mouse ES cells. The key steps are as follows:

1. Plate 5×10^4 undifferentiated ES cells on collagen IV coated plates in serum free DMEM/F12 supplemented with Activin A (20 ng/ml), bFGF (10 ng/ml), 1 % insulin-selenium-transferrin solution, 0.1 % BSA, 1 % MEM nonessential amino acids, 1 μM nucleotide mix, 0.1 mM β-mercaptoethanol, 2 mM glutamine, and 50 U/ml Pen/Strep.
2. Whole medium is changed on day 3 followed by changing half of the medium every day. During the incubation period the ES cells undergo a progressive change in the morphology and form isolated DE colonies by day 7 (Fig. 5a).
3. On day 9 the DE colonies are treated with TrypLE Express (Invitrogen, Carlsbad, CA) to obtain a single cell suspension. The typical yield is approximately 1×10^7 cells.
4. The DE cells express CXCR4 on the cell surface and therefore can be easily purified to >99 % purity by magnetic cell sorting using a two step labeling and magnetic cell separation: (a) label the DE cells in single cell suspension using PE conjugated anti-CXCR4 antibody (Miltenyi Biotec GmbH, Germany), (b) incubate the cells with anti-PE microbeads, (c) pass anti-PE microbead-labeled cells through the magnetic cell separation column and confirm their purity by flow cytometry (Fig. 5b).
5. These $CXCR4^+$ DE cells can either be transplanted under the kidney capsule or systemically injected in diabetic mice to spontaneously generate the insulin producing cells [40] (*see* **Notes 14–16**).

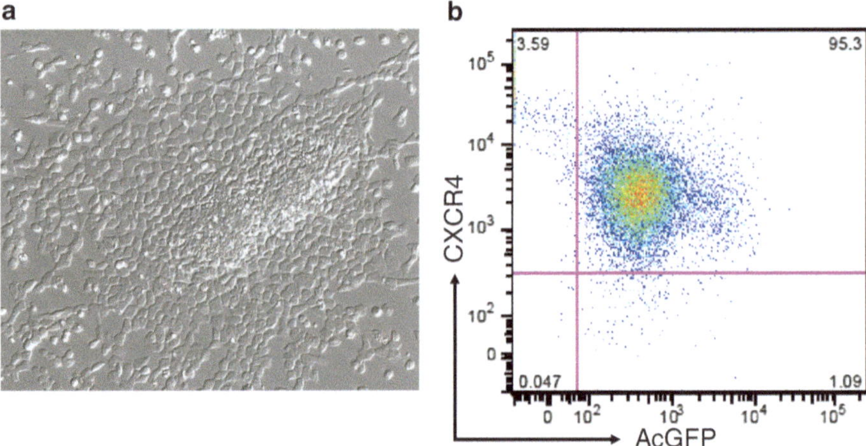

Fig. 5 Generation and characterization of DE cells: (**a**) R1 Pdx1 ES cells treated with recombinant Activin A for 9 days form a DE cell colony. (**b**) DE cells were analyzed for CXCR4 expression after labeling them with anti CXCR4 antibody and separating them with magnetic beads

4 Notes

1. Recent studies have demonstrated that the transgenes delivered by lentiviral vectors undergo silencing in ES cells in a promoter dependent manner [46, 47]. In these studies both EF1α and PGK promoters were shown to be capable of driving high levels of transgene expression. Taking advantage of this fact, we have now generated a stable Pdx1 expressing ES cell line R1Pdx1AcGFP/RIP-Luc [40].
2. The packaging plasmid pCFIVΔorf2Δvif contains full-length gag and pol, and rev, but has a deletion in the env gene, and mutations in vif and orf2 genes [44].
3. In a recent study, it has been shown that in comparison to human PAX4, mouse PAX4 displayed a tenfold increase in islet cell replication [45]. These results suggest that mouse Pax4 in contrast to human PAX4 may be more efficient in activating downstream target genes and enforcing a lineage commitment in the mouse ES cells towards pancreatic β-cell phenotype.
4. For achieving robust differentiation, the ES cells must be used at low passage and without primary mouse embryonic feeder cells.
5. It is best to use γ irradiated primary mouse embryonic feeder cells because residual mitomycin C may contribute to poor differentiation and even death of ES cells.
6. Continuous use of penicillin/streptomycin in the mouse ES cell medium can often mask a low level contamination with

mycoplasma. Therefore, it is highly recommended that the ES cell lines are regularly tested for mycoplasma contamination.

7. It is highly recommended that the ES cells are analyzed by karyotyping to rule out any chromosomal abnormalities especially at a higher passage number.
8. Periodically assess the pluripotency of the undifferentiated ES cells by performing flow cytometry following Oct3/4 and nanog staining.
9. Fluorescent reporters either fused in frame with the transgene or separated by the IRES or self-cleavable 2A peptide sequence allow for rapid detection of the transgene expression as well as permits FACS sorting for generating single ES cell clones.
10. Genetically engineered tagged recombinant proteins offer multiple advantages. Tags can facilitate easy and rapid detection of recombinant proteins during expression, simple one-step purification by affinity chromatography resulting in high purity, increased protein stability, solubility and allow on-column refolding. Detailed protocols for purification of proteins from baculovirus infected cells are well established [48].
11. Histidine-tags are small and therefore less disruptive to the properties of the proteins on which they are attached as a result it is not essential to remove the tag.
12. Imidazole is commonly used to elute His-tagged proteins. To minimize the binding of the unwanted host cell proteins, imidazole is used at a low concentration in the binding buffer. The recommended imidazole concentrations are 20 mM in the binding buffer and 500 mM in the elution buffer.
13. The PD10 desalting columns are prepacked, disposable columns containing Sephadex G25 for group separation of high ($M_r > 5{,}000$) from low ($M_r < 1{,}000$) molecular weight substances by desalting and buffer exchange. The protein yield using PD10 desalting columns is typically >95 % with <4 % salt contamination.
14. Tissue specific promoter driven luciferase reporter can be engineered in the ES cells. This unique strategy allows precise monitoring of the in vitro differentiation events that are difficult and cumbersome to monitor using traditional approaches. An additional benefit of this strategy is that tissue specific luciferase expression can be very efficiently used to monitor the fate and function of the transplanted cells by real-time noninvasive in vivo bioluminescence imaging [40, 49].
15. Thorel et al. [50] have described the conversion of adult pancreatic α-cells to β-cells following spontaneous *in vivo* reprogramming. This strategy has the potential for the generation

of higher numbers of pancreatic β cells directly from adult exocrine cells.

16. In a recent study, injection of rat wild-type pluripotent stem cells into $Pdx1^{-/-}$ mouse blastocysts led to the generation of normally functioning rat pancreas in $Pdx1^{-/-}$ mice [51].

Acknowledgements

This study was made possible by Grant Number 1I01BX001125-02, NIH/NHLBI #R01 HLO73015 and a VA Merit Review Award to N.Z. and by an NIH Pilot Grant to S.R. from the University of Iowa, Center for Gene Therapy of Cystic Fibrosis and Other Genetic Diseases, Grant Number NIH/NIDDK#P30 DK054759. S.R. is a recipient of the 2009 Young Investigator Award by the American Society of Transplantation.

References

1. Daneman D (2006) Type 1 diabetes. Lancet 367:847–858
2. Eisenbarth GS (1986) Type I diabetes mellitus. A chronic autoimmune disease. N Engl J Med 314:1360–1368
3. Shapiro AM, Lakey JR, Ryan EA, Korbutt GS, Toth E, Warnock GL, Kneteman NM, Rajotte RV (2000) Islet transplantation in seven patients with type 1 diabetes mellitus using a glucocorticoid-free immunosuppressive regimen. N Engl J Med 343:230–238
4. Shapiro AM, Ricordi C, Hering BJ, Auchincloss H, Lindblad R, Robertson RP, Secchi A, Brendel MD, Berney T, Brennan DC, Cagliero E, Alejandro R, Ryan EA, DiMercurio B, Morel P, Polonsky KS, Reems JA, Bretzel RG, Bertuzzi F, Froud T, Kandaswamy R, Sutherland DE, Eisenbarth G, Segal M, Preiksaitis J, Korbutt GS, Barton FB, Viviano L, Seyfert-Margolis V, Bluestone J, Lakey JR (2006) International trial of the Edmonton protocol for islet transplantation. N Engl J Med 355:1318–1330
5. Nathan DM (1993) Long-term complications of diabetes mellitus. N Engl J Med 328: 1676–1685
6. Nathan DM (2007) Finding new treatments for diabetes – how many, how fast... how good? N Engl J Med 356:437–440
7. Ricordi C, Strom TB (2004) Clinical islet transplantation: advances and immunological challenges. Nat Rev Immunol 4:259–268
8. Ryan EA, Lakey JR, Rajotte RV, Korbutt GS, Kin T, Imes S, Rabinovitch A, Elliott JF, Bigam D, Kneteman NM, Warnock GL, Larsen I, Shapiro AM (2001) Clinical outcomes and insulin secretion after islet transplantation with the Edmonton protocol. Diabetes 50:710–719
9. Ryan EA, Paty BW, Senior PA, Bigam D, Alfadhli E, Kneteman NM, Lakey JR, Shapiro AM (2005) Five-year follow-up after clinical islet transplantation. Diabetes 54:2060–2069
10. Smith RN, Kent SC, Nagle J, Selig M, Iafrate AJ, Najafian N, Hafler DA, Auchincloss H, Orban T, Cagliero E (2008) Pathology of an islet transplant 2 years after transplantation: evidence for a nonimmunological loss. Transplantation 86:54–62
11. Evans MJ, Kaufman MH (1981) Establishment in culture of pluripotential cells from mouse embryos. Nature 292:154–156
12. Takahashi K, Yamanaka S (2006) Induction of pluripotent stem cells from mouse embryonic and adult fibroblast cultures by defined factors. Cell 126:663–676
13. Takahashi K, Tanabe K, Ohnuki M, Narita M, Ichisaka T, Tomoda K, Yamanaka S (2007) Induction of pluripotent stem cells from adult human fibroblasts by defined factors. Cell 131:861–872
14. Thomson JA, Itskovitz-Eldor J, Shapiro SS, Waknitz MA, Swiergiel JJ, Marshall VS, Jones JM (1998) Embryonic stem cell lines derived from human blastocysts. Science 282: 1145–1147
15. Lumelsky N, Blondel O, Laeng P, Velasco I, Ravin R, McKay R (2001) Differentiation of

embryonic stem cells to insulin-secreting structures similar to pancreatic islets. Science 292:1389–1394

16. Blyszczuk P, Czyz J, Kania G, Wagner M, Roll U, St-Onge L, Wobus AM (2003) Expression of Pax4 in embryonic stem cells promotes differentiation of nestin-positive progenitor and insulin-producing cells. Proc Natl Acad Sci U S A 100:998–1003
17. Hori Y, Rulifson IC, Tsai BC, Heit JJ, Cahoy JD, Kim SK (2002) Growth inhibitors promote differentiation of insulin-producing tissue from embryonic stem cells. Proc Natl Acad Sci U S A 99:16105–16110
18. Miyazaki S, Yamato E, Miyazaki J (2004) Regulated expression of pdx-1 promotes in vitro differentiation of insulin-producing cells from embryonic stem cells. Diabetes 53: 1030–1037
19. Shi Y, Hou L, Tang F, Jiang W, Wang P, Ding M, Deng H (2005) Inducing embryonic stem cells to differentiate into pancreatic beta cells by a novel three-step approach with activin A and all-trans retinoic acid. Stem Cells 23:656–662
20. Soria B, Roche E, Berna G, Leon-Quinto T, Reig JA, Martin F (2000) Insulin-secreting cells derived from embryonic stem cells normalize glycemia in streptozotocin-induced diabetic mice. Diabetes 49:157–162
21. Treff NR, Vincent RK, Budde ML, Browning VL, Magliocca JF, Kapur V, Odorico JS (2006) Differentiation of embryonic stem cells conditionally expressing neurogenin 3. Stem Cells 24:2529–2537
22. Ahlgren U, Jonsson J, Edlund H (1996) The morphogenesis of the pancreatic mesenchyme is uncoupled from that of the pancreatic epithelium in IPF1/PDX1-deficient mice. Development 122:1409–1416
23. Holland AM, Gonez LJ, Naselli G, MacDonald RJ, Harrison LC (2005) Conditional expression demonstrates the role of the homeodomain transcription factor Pdx1 in maintenance and regeneration of beta-cells in the adult pancreas. Diabetes 54:2586–2595
24. Offield MF, Jetton TL, Labosky PA, Ray M, Stein RW, Magnuson MA, Hogan BL, Wright CV (1996) PDX-1 is required for pancreatic outgrowth and differentiation of the rostral duodenum. Development 122: 983–995
25. Stoffers DA, Heller RS, Miller CP, Habener JF (1999) Developmental expression of the homeodomain protein IDX-1 in mice transgenic for an IDX-1 promoter/lacZ transcriptional reporter. Endocrinology 140:5374–5381
26. Ferber S, Halkin A, Cohen H, Ber I, Einav Y, Goldberg I, Barshack I, Seijffers R, Kopolovic J, Kaiser N, Karasik A (2000) Pancreatic and duodenal homeobox gene 1 induces expression of insulin genes in liver and ameliorates streptozotocin-induced hyperglycemia. Nat Med 6:568–572
27. Heller RS, Stoffers DA, Bock T, Svenstrup K, Jensen J, Horn T, Miller CP, Habener JF, Madsen OD, Serup P (2001) Improved glucose tolerance and acinar dysmorphogenesis by targeted expression of transcription factor PDX-1 to the exocrine pancreas. Diabetes 50:1553–1561
28. Wang H, Maechler P, Ritz-Laser B, Hagenfeldt KA, Ishihara H, Philippe J, Wollheim CB (2001) Pdx1 level defines pancreatic gene expression pattern and cell lineage differentiation. J Biol Chem 276:25279–25286
29. Ahlgren U, Jonsson J, Jonsson L, Simu K, Edlund H (1998) Beta-cell-specific inactivation of the mouse Ipf1/Pdx1 gene results in loss of the beta-cell phenotype and maturity onset diabetes. Genes Dev 12:1763–1768
30. Brissova M, Shiota M, Nicholson WE, Gannon M, Knobel SM, Piston DW, Wright CV, Powers AC (2002) Reduction in pancreatic transcription factor PDX-1 impairs glucose-stimulated insulin secretion. J Biol Chem 277:11225–11232
31. Kulkarni RN, Jhala US, Winnay JN, Krajewski S, Montminy M, Kahn CR (2004) PDX-1 haploinsufficiency limits the compensatory islet hyperplasia that occurs in response to insulin resistance. J Clin Invest 114:828–836
32. Jonsson J, Carlsson L, Edlund T, Edlund H (1994) Insulin-promoter-factor 1 is required for pancreas development in mice. Nature 371:606–609
33. Collombat P, Xu X, Ravassard P, Sosa-Pineda B, Dussaud S, Billestrup N, Madsen OD, Serup P, Heimberg H, Mansouri A (2009) The ectopic expression of Pax4 in the mouse pancreas converts progenitor cells into alpha and subsequently beta cells. Cell 138:449–462
34. Smith SB, Qu HQ, Taleb N, Kishimoto NY, Scheel DW, Lu Y, Patch AM, Grabs R, Wang J, Lynn FC, Miyatsuka T, Mitchell J, Seerke R, Desir J, Eijnden SV, Abramowicz M, Kacet N, Weill J, Renard ME, Gentile M, Hansen I, Dewar K, Hattersley AT, Wang R, Wilson ME, Johnson JD, Polychronakos C, German MS (2010) Rfx6 directs islet formation and insulin production in mice and humans. Nature 463:775–780
35. Sosa-Pineda B, Chowdhury K, Torres M, Oliver G, Gruss P (1997) The Pax4 gene is essential for differentiation of insulin-producing beta cells in the mammalian pancreas. Nature 386:399–402
36. Wang J, Elghazi L, Parker SE, Kizilocak H, Asano M, Sussel L, Sosa-Pineda B (2004) The concerted activities of Pax4 and Nkx2.2 are

essential to initiate pancreatic beta-cell differentiation. Dev Biol 266:178–189
37. Alipio Z, Liao W, Roemer EJ, Waner M, Fink LM, Ward DC, Ma Y (2010) Reversal of hyperglycemia in diabetic mouse models using induced-pluripotent stem (iPS)-derived pancreatic beta-like cells. Proc Natl Acad Sci U S A 107:13426–13431
38. Bernardo AS, Cho H, Mason S, Docherty HM, Pedersen RA, Vallier L, Docherty K (2008) Biphasic induction of Pdx1 in mouse and human embryonic stem cells can mimic development of pancreatic beta cells. Stem Cells 27:341–351
39. Oliver-Krasinski JM, Kasner MT, Yang J, Crutchlow MF, Rustgi AK, Kaestner KH, Stoffers DA (2009) The diabetes gene Pdx1 regulates the transcriptional network of pancreatic endocrine progenitor cells in mice. J Clin Invest 119:1888–1898
40. Raikwar S, Zavazava N (2010) Spontaneous in vivo differentiation of ES cell-derived pancreatic endoderm-like cells into insulin producing cells corrects hyperglycemia in diabetic mice. Transplantation 91:11–20
41. Raikwar SP, Zavazava N (2009) Insulin producing cells derived from embryonic stem cells: are we there yet? J Cell Physiol 218:256–263
42. Shiraki N, Yoshida T, Araki K, Umezawa A, Higuchi Y, Goto H, Kume K, Kume S (2008) Guided differentiation of embryonic stem cells into Pdx1-expressing regional-specific definitive endoderm. Stem Cells 26:874–885
43. Kost TA, Condreay JP, Jarvis DL (2005) Baculovirus as versatile vectors for protein expression in insect and mammalian cells. Nat Biotechnol 23:567–575
44. Stein CS, Davidson BL (2002) Gene transfer to the brain using feline immunodeficiency virus-based lentivirus vectors. Methods Enzymol 346:433–454
45. Brun T, Hu He KH, Lupi R, Boehm B, Wojtusciszyn A, Sauter N, Donath M, Marchetti P, Maedler K, Gauthier BR (2007) The diabetes-linked transcription factor Pax4 is expressed in human pancreatic islets and is activated by mitogens and GLP-1. Hum Mol Genet 17:478–489
46. Kim S, Kim GJ, Miyoshi H, Moon SH, Ahn SE, Lee JH, Lee HJ, Cha KY, Chung HM (2007) Efficiency of the elongation factor-1alpha promoter in mammalian embryonic stem cells using lentiviral gene delivery systems. Stem Cells Dev 16:537–545
47. Xia X, Zhang Y, Zieth CR, Zhang SC (2007) Transgenes delivered by lentiviral vector are suppressed in human embryonic stem cells in a promoter-dependent manner. Stem Cells Dev 16:167–176
48. O'Shaughnessy L, Doyle S (2011) Purification of proteins from baculovirus-infected insect cells. Methods Mol Biol 681:295–309
49. Raikwar SP, Zavazava N (2009) Real-time non-invasive imaging of ES cell-derived insulin producing cells. Methods Mol Biol 590: 317–334
50. Thorel F, Nepote V, Avril I, Kohno K, Desgraz R, Chera S, Herrera PL (2010) Conversion of adult pancreatic alpha-cells to beta-cells after extreme beta-cell loss. Nature 464: 1149–1154
51. Kobayashi T, Yamaguchi T, Hamanaka S, Kato-Itoh M, Yamazaki Y, Ibata M, Sato H, Lee YS, Usui J, Knisely AS, Hirabayashi M, Nakauchi H (2010) Generation of rat pancreas in mouse by interspecific blastocyst injection of pluripotent stem cells. Cell 142: 787–799

Chapter 8

Mouse ES Cell-Derived Hematopoietic Progenitor Cells

Eun-Mi Kim, Gohar Manzar, and Nicholas Zavazava

Abstract

Future stem cell-based therapies will benefit from the new discoveries being made on pluripotent stem cells such as embryonic stem (ES) cells and induced pluripotent stem (IPS) cells. Understanding the genes regulating pluripotency has opened new opportunities to generate patient-tailored therapies. However, protocols for deriving progenitor cells of therapeutic grade from these pluripotent stem cells are not yet worked out. In particular the potential of these cells in treating diseases when compared to their adult progenitor counterparts is unknown. This is crucial work that needs to be studied in detail because we will need to determine engraftment potential of these cells and their ability for multi-lineage engraftment in the in vivo setting before any clinical applications. The ability of these cells to engraft is dependent on their expression of cell surface markers which guide their homing patterns. In this review, I discuss murine hematopoietic progenitor cells derived from mouse ES cells. Stem cells in the bone marrow are found in the bone marrow niches. Our knowledge of the bone marrow niches is growing and will ultimately lead to improved clinical transplantation of bone marrow cells. We are, however, a long way in appreciating how hematopoietic progenitor cells migrate and populate lymphoid tissues. One of the variables in generating hematopoietic progenitor cells is that different labs use different approaches in generating progenitor cells. In some cases, the ES cell lines used show some variability as well. The cell culture media used by the different investigators highly influence the maturation level of the cells and their homing patterns. Here, mouse ES cell-derived progenitor cells are discussed.

Key words HoxB4, Embryonic stem cells, NK cells, Stemness, Ectoderm, Mesoderm, Endoderm

1 Introduction

Bone marrow stem cells that are capable of replenishing peripheral blood cells are defined as Lin$^-$Sca-1^{high}c-kit^{high} or otherwise known as LSK cells [1]. These cells have now been established to contain the "stemness" of bone marrow cells. One important quality of these cells is that they are capable of symmetric and asymmetric cell division within the bone marrow niche. So far, at least two bone marrow niches have been defined, the endosteal and the perivascular niches, respectively [2]. It is not clear whether these two bone marrow niches are the end of the story or whether new niches will be discovered in the future as investigators seek to better

Nicholas Zavazava (ed.), *Embryonic Stem Cell Immunobiology: Methods and Protocols*, Methods in Molecular Biology, vol. 1029, DOI 10.1007/978-1-62703-478-4_8, © Springer Science+Business Media New York 2013

understand the bone marrow niches. The niches are not unique to the bone marrow, but rather are thought to be present throughout the body, particularly in those organs capable of regeneration such as the liver, gastrointestinal tract, the skin and blood. The niche is responsible for tissue self renewal as new progenitor cells are released as a response to stimuli from tissue injuries. These enter the differentiation phase to generate adult tissue. Pluripotent stem cell-generated progenitor cells must possess these qualities to be of any biological significance.

For the progenitor cells to be able to enter the bone marrow niches, they need to express certain chemokine receptors. For example CXCL12, CXCL9, stem cell factor, and a few others appear to be key players in enabling bone marrow progenitor cells to home into the niches. It has not been easy to generate definitive hematopoietic cells in both mouse and human pluripotent stem cells. For example, we published a manuscript a few years ago in which we reported that ES cell derived mouse ES cells failed to engraft permanently [3]. Although we were able to generate progenitor cells that expressed hematopoietic cell surface markers, the cells initially induced mixed chimerism in immunodeficient and immunocompetent mice but were no longer detectable after 3–4 weeks in peripheral blood. We concluded from these experiments that the progenitor cells did not have self-renewal capabilities.

The discovery of HoxB4 as a gene in a leukemia that provides self-renewal to cancer cells provided a possible answer as to how these cells can be made to survive long-term. It had been shown that transduction of HoxB4 in hematopoietic cells led to the expansion of progenitor cells in vitro and in vivo [4, 5]. This transcription factor had the same function in both mouse and human hematopoietic cells. Transduction of HoxB4 in mouse ES cells allowed the generation of hematopoietic progenitor cells that had long-term survival [6]. Although the hematopoietic stem cells derived using this approach could engraft indefinitely, there was a problem with lineage commitment. Most of the cells derived using this approach were myeloid and the conclusion by these authors and others was that HoxB4 impeded lymphocytic cell development in vitro and in vivo. However, long-term engraftment was achieved. Unfortunately, the cells were virally transduced to avoid gene silencing by the use of plasmids in ES cells. Although no tumors have been so far described in mice transplanted with HoxB4-transduced cells, another group observed tumor formation in dogs transplanted with HoxB4-transduced hematopoietic cells [7]. An additional problem is the lack of lymphocytes in hematopoietic cells transduced with HoxB4. These two disadvantages are an impediment to studying ES cell-derived progenitor cells and their possibility to restore immunity in immunodeficient mice. Despite these problems, HoxB4 allows the study of the progenitor cells derived from mouse ES cells. Thus, we can determine the heterogeneity of these cells and their expression of cell markers.

2 Generation of Mouse Hematopoietic Progenitor Cells

Pluripotent stem cells are unique in their ability to form embryoid bodies (EBs). These EBs are cell clusters that represent a mixture of cells from all three germ layers, the mesoderm, ectoderm and endoderm. The ability of cells to form EBs is the mainstay for generating hematopoietic cells from ES cells but also for confirming pluripotency of ES cells or that of iPS cells. EBs can be generated by different techniques, the hanging drop technique, by methycellulose culture or by suspension culture [8]. Initially, these EBs contain erythroid precursors by day 4. By day 6 higher cell numbers of hematopoietic cells can be harvested from the EBs. In some reports addition of BMP4 and VEGF enhances hematopoietic cell development [9, 10]. The EBs are generally dismantled and the free cell suspension treated with a cocktail of cytokines. This is followed by a massive cell death of non-mesodermal cells which can be sorted out. Early markers of hematopoietic cells are FLK1, CD31, CD41 and finally CD45. Regular tracking of these molecules during the differentiation process allows easy monitoring of hematopoiesis [6, 11, 12]. Figure 1 shows the embryoid body protocol for generating hematopoietic progenitor cells.

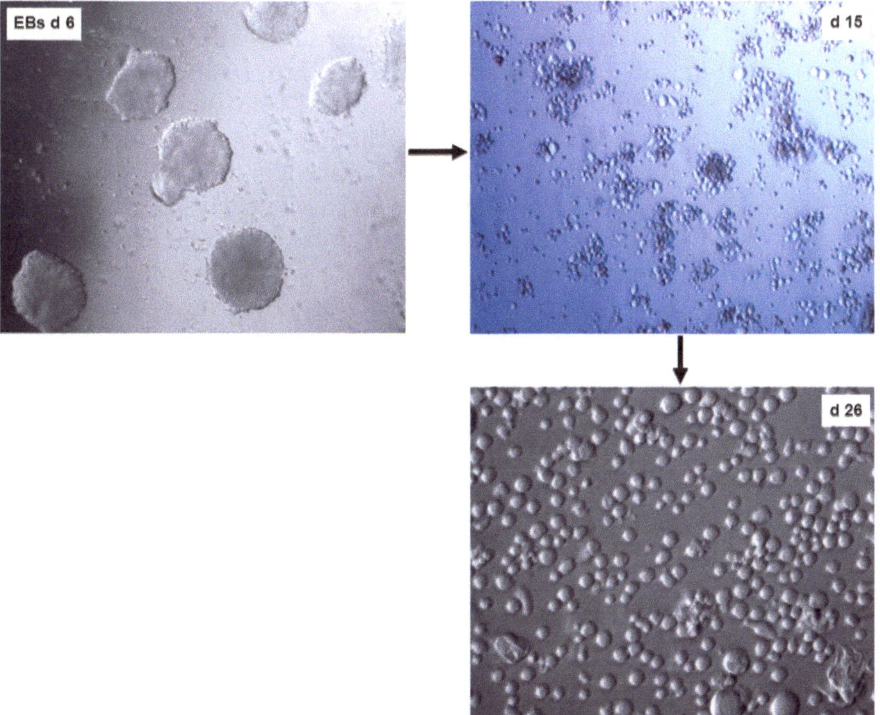

Fig. 1 Progenitors of pluripotent stem cells. iPS or ES cells can be differentiated into multiple cell lineages, such as insulin producing cells, neuronal cells or hematopoietic cells. Studies have been carried out on the progenitors to determine the impact of T cells, NK cells but unfortunately no studies have yet been performed on the humoral responses to these cells

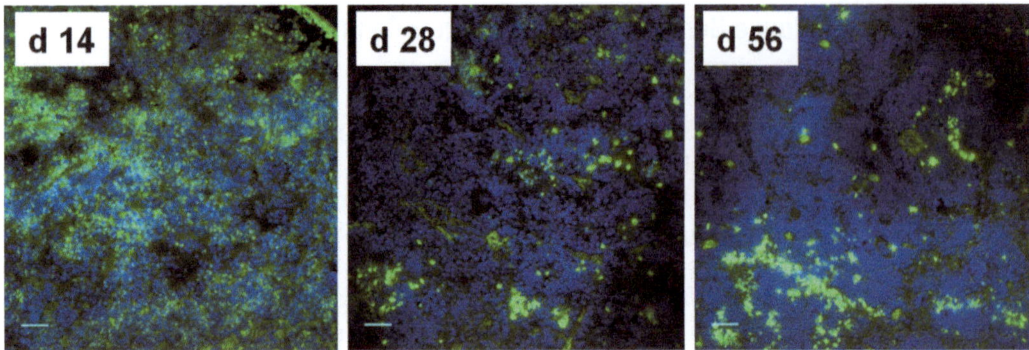

Fig. 2 ES cell-derived hematopoietic cells populate peripheral lymphoid organs. ES cells were differentiated into HPCs. These cells were transplanted into Rag2$^{-/-}\gamma c^{-/-}$ mice. Mice were sacrificed after 14, 28 or 56 days. In the spleen, HPCs were consistently detected. The cell numbers decreased with time

When transplanted in immunodeficient mice, these cells populate lymphoid tissues and actively divide replenishing hematopoietic cells, Fig. 2.

A more efficient method for generating hematopoietic cells from ES cells is the use of stromal cell lines. The OP9 is a mesenchymal cell line that expresses only class I antigens, but no class II molecules. It was described by Nagano et al. [13] and has been used for generating hematopoietic cells from both mouse and human cells [14, 15]. Another stromal cell line, the MS-5 together with growth factors KL, IL-3, IL-6, IL-11, G-CSF, and Epo seem to enhance hematopoiesis [16, 17]. Megakaryocytes were formed after addition of TPO. The generation of lymphocytes from ES cells has been much more challenging. On OP9 cells, B cells, erythroid and myeloid cells were generated [17]. OP9 stromal cells produce SCF and IL-7, which are important for hematopoietic cell development. The addition of Flt3 led to a tenfold increase in B-cell production with reduced erythroid and myeloid differentiation.

The generation of T, B and NK cells requires Notch-1 signaling. Zuniga-Pflucker et al. described a protocol which requires an initial culture of ES cells on OP9 cells up to day 8. The cells at this stage have developed into CD41- and CD45-expressing cells. Subsequently, the hematopoietic progenitor cells are transferred onto OP cells that express DL-1, a ligand for Notch-1 [15, 18]. The cells undergo a series of steps from the DN1 phase through the CD4$^+$CD8$^+$ stage. When transplanted in autologous mice these cells induce lethal graft versus disease [1]. Interestingly, these T cells become single CD8 positive but not CD4 positive. The lack of class II expression by the OP9 stromal cells is thought to be the reason why no CD4$^+$ T cells develop in these cultures. To avoid GVHD, the T cell precursor cells can be transferred onto fetal organ cultures during the DN1 and DN2 phases. Thymic selection allows the derivation of T cells tolerant to self but that are responsive to viral infection [18].

Interestingly, these progenitor cells are not capable of long-term survival in the mouse. However, transduction of ES cells with HoxB4 confers self-renewal properties allowing long-term survival. This is important as it allows long-term studies to be carried out such as the study of T cell memory. Earlier studies had claimed that HoxB4 blocks T cell development [6]. Our own studies seem to suggest that the lack of robust lymphocyte-development in those earlier studies might have been due to the particular protocols used. For example, without Notch-1 signaling, T cell development is impaired.

3 Long Term Survival of Hematopoietic Progenitor Cells

Many of the data discussed so far showed lack of long-term engraftment of ES cell-derived hematopoietic progenitor cells post transplantation in immunodeficient mice. Our own experiments showed that after transplantation of non-differentiated ES cells, modest mixed chimerism was established but this was accompanied by massive apoptosis in lymphoid tissues [3]. The results of these experiments were very much similar to data obtained after transplanting ES cell-derived progenitor cells. This indicated that ES cell-derived hematopoietic progenitor cells lacked self-renewal properties because the result was the same in both immunodeficient mice and in allogeneic mice. HoxB4 is expressed in the yolk sack at the initiation of hematopoiesis [19]. Sauvageau et al. [20] reported that overexpression of HoxB4 in hematopoietic cells led to the in vitro and in vivo expansion of primitive hematopoietic cells. On the other hand, overexpression of HoxB4 in ES cells enhanced the erythropoietic, and possibly more primitive, hematopoietic differentiation potential of ES cells. In those experiments, the investigators concluded that there was no effect on myeloid cells. Results by Kyba et al. [6] a few years later however showed that hematopoietic progenitor cells derived from HoxB4-transduced ES cells preferentially differentiated into myeloid cells rather than into lymphoid cells. Indeed this has been confirmed by others and by our own studies [11, 12]. So far no data have yet been reported on HoxB4-transduced cells being able to differentiate into lymphoid cells. Our lab has most recently successfully derived T cells from HoxB4-transduced ES cells, showing that if the right cues are delivered, the suggested "repressive effect" of HoxB4 on lymphopoiesis can be abolished [1]. The expression of key T cell transcription factors Notch-1, Skp2, or E47 were not perturbed in these cells. Thus, a better understanding of the signaling cues in the HoxB4-transduced ES cells would allow improved study of whether ES cell-derived cells can fully reconstitute the immune system.

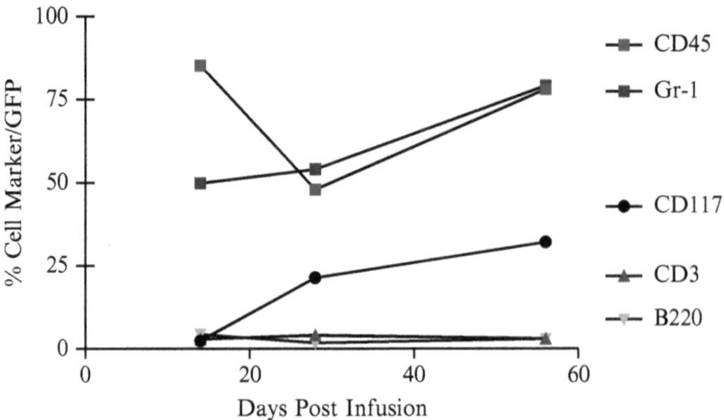

Fig. 3 ES cell-derived HPCs form multi-lineage chimerism. 129SvJ ES cell-derived HPCs were transplanted in 129SvJ mice and multi-lineage chimerism monitored using flow cytometry. All major cell lineages were detectable in peripheral blood up to 56 days, suggesting that ES cell-derived HPCs are capable of reconstituting peripheral blood

One of the problems in the use of HoxB4 is that permanent ectopic expression seems to compromise the ability of the hematopoietic progenitor cells to go into multi-lineage hematopoiesis and might interfere with the engraftment of the cells. Another approach that has been used is the combination of HoxB4 and Cdx4. Ectopic expression of Cdx4 under a tetracycline-inducible system increased mesoderm specification, blast colony and hematopoietic progenitor formation. Together with HoxB4, enhanced long-term multilineage engraftment was observed [21]. Both CD4 and CD8 positive cells were observed and normal thymic repopulation developed. However, these gene systems appear quite involved and difficult for smaller labs to recapitulate. Further, it has been reported that the amount of HoxB4 expressed in the transduced cells affects hematopoietic cell development [22]. Despite these caveats, HoxB4 has allowed a better study of ES cell-derived HPCs and their ability to populate bone marrow spaces. This would have been impossible with non-HoxB4 transduced ES cells as they are not capable of self-renewal, which only engraft for a very limited time of up to 4 weeks in vivo. Although a few cells may persist in peripheral blood, the pool size of these cells is too small to have any impact on immunity, Fig. 3.

4 Functionality of ES Cell-Derived HPCs

Data on functional studies of ES cell-derived HPCs are very limited in the literature. Yet this is an important aspect in determining whether ES cells can become an alternative source of hematopoietic cells someday. Schmitt et al. [15] functionally analyzed T cells derived from mouse ES cells using the OP9/OP9-DL1 stromal cells. First, they stimulated the T cells with a combination of plate bound anti-CD3 and and-CD28 antibodies and successfully showed that the T cells responded by producing IFN-γ. Immmunophenotyping of the cells also revealed that the cells had a normal T cell repertoire. Finally, infection of immunodeficient mice reconstituted with ES cell derived T cells revealed a normal T cell response to viral antigen. These results showed that ES cell-derived T cells indeed respond to antigen. The caveat is that these experiments were short-term experiments completed within the first 3 weeks of reconstituting mice. It remains to be seen whether the T cells derived from HoxB4-transduced ES cells are capable of long term survival and are capable of developing T cell memory.

Our initial experiments showed that mice reconstituted with ES cell-derived HPCs induced mixed chimerism including that of donor T cells. When these mice were infected with the LCMV virus, they also responded by developing antigen-specific CD8 T cells [3]. More recently, using the OP9 stromal cell lines, we successfully showed that in vitro-generated T cells that were not generated through the use of fetal organ cultures led to severe and fatal graft-versus-host reactions in autologous recipient mice. When the cells were cultured on fetal organ cultures, however, negative selection led to the generation of immunocompetent T cells that were responsive to antigen stimulation but tolerant to self [1]. Our results confirmed that ES cell-derived T cells are immunocompetent. Thus, indeed HoxB4 does not appear to repress the development of T cells, but rather confers self-renewal properties.

The generation of NK cells from ES cells has also been successful. Lian et al. [23] reported a protocol that allowed the generation of NK cells from mouse ES cells. These NK cells expressed NK cell receptors and efficiently lysed target cells that poorly expressed MHC antigens. More recently, this has been also possible with human ES cells [24, 25] which have been used to derive NK cells that effectively lyse tumor cells. These data are significant in that they show that indeed human ES cells can be used to derive lineage-specific cells that potentially can be used in the treatment of human disease.

However, in the generation of B cells, the data are still unclear. For example, Martin et al. [26] failed to derive T or B cells from human ES cells but were readily successful at deriving these cells from umbilical $CD34^+$ cells. This is a major limitation to the use of human ES cells in producing cells that are directed at human therapies. In contrast, Carpenter et al. successfully generated $CD45^+CD10^+$ cells that were positive for transcripts Pax5, IL7αR, lambda like, and VpreB receptor. The cells were, however, negative for surface IgM and CD5 expression, iPS derived $CD45^+CD19^+$ cells also exhibited multiple genomic D-J(H) rearrangements, which supports a pre-B cell identity. So far though, no data are available in the human system where mature functional B cells have been described. These data show that there are clear differences in gene regulation between human and mouse pluripotent stem cells. T and B cells have been derived from mouse ES cells, but this seems a lot more challenging in human pluripotent stem cells.

5 Conclusions

While the derivation of hematopoietic cells from both murine and human pluripotent stem cells is improving by the day, it has so far not been possible to establish long-term engraftment of human hematopoietic progenitor cells. $CD34^+$ progenitors can be generated readily; engraftment studies have only shown short-term survival of these cells in mice at a low degree of chimerism [2]. This finding is similar to that found in mouse progenitor cells. The results seem to suggest that survival transcription factors have not been turned on in these progenitor cells. Although HoxB4 has been found to confer self-renewal in mouse cells, this does not appear to be the case with human cells. In fact the data so far are conflicting [3, 4]. Therefore, the process of self-renewal in human hematopoietic cells is differentially regulated. Till we establish a protocol that provides the necessary maturation and self-renewal of human $CD34^+$ cells, it will remain difficult to replicate mouse studies in humans.

6 Acknowledgments

This material is based upon work supported in part by the Department of Veterans Affairs, Veterans Health Administration, Office of Research and Development, Biomedical Laboratory Research and Development 1 I01 BX001125-02 and by NIH/NHLBI Grants 5 R01 HL073015-08 and 3 R01 HL073015-04A1S1. The content of this manuscript is solely the responsibility of the authors and does not necessarily represent the official views of the granting agencies. The authors have no conflicts of interest to report.

References

1. Henckaerts E, Langer JC, Orenstein J, Snoeck HW (2004) The positive regulatory effect of TGF-beta2 on primitive murine hemopoietic stem and progenitor cells is dependent on age, genetic background, and serum factors. J Immunol 173:2486–2493
2. Ehninger A, Trumpp A (2011) The bone marroe stem cell niche grows up: mesenchymal stem cells and macrophages move in. J Exp Med 208:421–428, Ref Type: Generic
3. Bonde S, Zavazava N (2006) Immunogenicity and engraftment of mouse embryonic stem cells in allogeneic recipients. Stem Cells 24: 2192–2201
4. Helgason CD, Sauvageau G, Lawrence HJ, Largman C, Humphries RK (1996) Overexpression of HOXB4 enhances the hematopoietic potential of embryonic stem cells differentiated in vitro. Blood 87: 2740–2749

5. Thorsteinsdottir U, Sauvageau G, Humphries RK (1999) Enhanced in vivo regenerative potential of HOXB4-transduced hematopoietic stem cells with regulation of their pool size. Blood 94:2605–2612
6. Kyba M, Perlingeiro RCR, Daley GQ (2002) HoxB4 confers definitive lymphoid-myeloid engraftment potential on embryonic stem cell and yolk sac hematopoietic progenitors. Cell 109:29–37
7. Zhang XB et al (2008) High incidence of leukemia in large animals after stem cell gene therapy with a HOXB4-expressing retroviral vector. J Clin Invest 118:1502–1510
8. Dang SM, Kyba M, Perlingeiro R, Daley GQ, Zandstra PW (2002) Efficiency of embryoid body formation and hematopoietic development from embryonic stem cells in different culture systems. Biotechnol Bioeng 78: 442–453
9. Nakayama N, Lee J, Chiu L (2000) Vascular endothelial growth factor synergistically enhances bone morphogenetic protein-4-dependent lymphohematopoietic cell generation from embryonic stem cells in vitro. Blood 95:2275–2283
10. Bonde S, Dowden AM, Chan KM, Tabayoyong WB, Zavazava N (2008) HOXB4 but not BMP4 confers self-renewal properties to ES-derived hematopoietic progenitor cells. Transplantation 86:1803–1809
11. Chan KM, Bonde S, Klump H, Zavazava N (2008) Hematopoiesis and immunity of HOXB4-transduced embryonic stem cell-derived hematopoietic progenitor cells. Blood 111:2953–2961
12. Pilat S et al (2005) HOXB4 enforces equivalent fates of ES-cell-derived and adult hematopoietic cells. Proc Natl Acad Sci U S A 102:12101–12106
13. Nakano T, Kodama H, Honjo T (1994) Generation of lymphohematopoietic cells from embryonic stem cells in culture. Science 265:1098–1101
14. Nakano T, Kodama H, Honjo T (1996) In vitro development of primitive and definitive erythrocytes from different precursors. Science 272:722–724
15. Schmitt TM et al (2004) Induction of T cell development and establishment of T cell competence from embryonic stem cells differentiated in vitro. Nat Immunol 5:410–417
16. Berthier R et al (1997) The MS-5 murine stromal cell line and hematopoietic growth factors synergize to support the megakaryocytic differentiation of embryonic stem cells. Exp Hematol 25:481–490
17. Mossuz P, Schweitzer A, Molla A, Berthier R (1997) Expression and function of receptors for extracellular matrix molecules in the differentiation of human megakaryocytes in vitro. Br J Haematol 98:819–827
18. Schmitt TM, Zuniga-Pflucker JC (2002) Induction of T cell development from hematopoietic progenitor cells by delta-like-1 in vitro. Immunity 17:749–756
19. McGrath KE, Palis J (1997) Expression of homeobox genes, including an insulin promoting factor, in the murine yolk sac at the time of hematopoietic initiation. Mol Reprod Dev 48:145–153
20. Sauvageau G et al (1995) Overexpression of HOXB4 in hematopoietic cells causes the selective expansion of more primitive populations in vitro and in vivo. Genes Dev 9:1753–1765
21. Wang Y, Yates F, Naveiras O, Ernst P, Daley GQ (2005) Embryonic stem cell-derived hematopoietic stem cells. Proc Natl Acad Sci U S A 102:19081–19086
22. Schiedlmeier B et al (2003) High-level ectopic HOXB4 expression confers a profound in vivo competitive growth advantage on human cord blood CD34+ cells, but impairs lymphomyeloid differentiation. Blood 101: 1759–1768
23. Lian RH et al (2002) Orderly and nonstochastic acquisition of CD94/NKG2 receptors by developing NK cells derived from embryonic stem cells in vitro. J Immunol 168: 4980–4987
24. Woll PS, Martin CH, Miller JS, Kaufman DS (2005) Human embryonic stem cell-derived NK cells acquire functional receptors and cytolytic activity. J Immunol 175:5095–5103
25. Wilber A et al (2007) Efficient and stable transgene expression in human embryonic stem cells using transposon-mediated gene transfer. Stem Cells 25:2919–2927
26. Martin CH, Woll PS, Ni Z, Zuniga-Pflucker JC, Kaufman DS (2008) Differences in lymphocyte developmental potential between human embryonic stem cell and umbilical cord blood-derived hematopoietic progenitor cells. Blood 112:2730–2737

Chapter 9

Directed Differentiation of Embryonic Stem Cells to the T-Lymphocyte Lineage

Haydn C.-Y. Liang, Roxanne Holmes, and Juan Carlos Zúñiga-Pflücker

Abstract

Hematopoiesis is the highly regulated and complex process by which blood cells are formed. Hematopoiesis can be achieved in vitro by the differentiation of embryonic stem cells (ESCs) into hematopoietic lineage cells. Differentiation of ESCs initially gives rise to mesoderm colonies that go on to form hemangioblast cells, which possess endothelial and hematopoietic lineage potential. While the differentiation of several hematopoietic lineages from ESCs, such as erythrocytes and macrophages, can be easily recapitulated in vitro, T-cell differentiation requires additional Notch-dependent signals for their generation. Keeping with this, ESCs induced to differentiate with OP-9 cells, a bone marrow-derived stromal cell line, give rise to erythro-myeloid cells and B lymphocytes, while the expression of an appropriate Notch ligand, such as Delta-like 1, on OP-9 cells (OP9-DL1) is required to support the generation of T-cells in vitro. Here, we describe an updated and streamlined protocol for the generation of T-lineage cells from mouse ESCs cultured on OP9-DL1 cells. This approach can facilitate studies aimed to assess the effects of environmental and genetic manipulations at various stages of T-cell development.

Key words OP9, OP9-DL, T lymphocytes, Macrophage colony-stimulating factor

1 Introduction

Embryonic stem cells (ESCs) are pluripotent cells derived from the inner cell mass of the blastocyst capable of differentiating into all cell types found in an adult organism, including hematopoietic cells [1]. While several approaches have been described for the in vitro differentiation of hematopoietic cells from ESCs [2–7], each method offers different advantages to aid in the characterization, manipulation, and analysis of ESC-derived lymphopoiesis. The ESC-OP9-DL1 coculture system described here was initially modified from the seminal work by Nakano et al. in which B-cell differentiation was obtained in vitro from cocultures of ESCs and the bone marrow stromal cell line, OP-9, isolated from an *op*/*op* mouse [6–8]. OP-9 cells lack the expression of macrophage colony-stimulating factor (M-CSF), which prevents the cellular expansion

Nicholas Zavazava (ed.), *Embryonic Stem Cell Immunobiology: Methods and Protocols*, Methods in Molecular Biology, vol. 1029, DOI 10.1007/978-1-62703-478-4_9,

of macrophages, allowing other hematopoietic lineages such as B-cells to thrive in culture [8]. Further refinement of the initial protocol, including the provision of additional IL-7 and Flt-3L to cocultures [9–15], resulted in improved B-cell production [9]. This established an efficient and robust model for ESC to B-cell development in vitro [9], but T-cells were not produced.

Notch signaling was shown to be a critical event for T-cell development [16, 17]. By retrovirally transducing OP-9 cells to express the Notch ligand Delta-like 1 (OP9-DL1), it was shown that T-cell, instead of B-cell, development from stem cells takes place under similar conditions [18–21]. This finding has been demonstrated with ESCs and other progenitors, such as fetal liver hematopoietic stem cells (HSCs), adult bone marrow HSCs, and ESC-derived progenitors.

Here, we describe an updated and streamlined protocol for the generation of T-lineage cells from mouse ESCs cultured on OP9-DL1 cells. The method described here provides a robust coculture system for T-cell development from ESCs up to the CD8 $\alpha\beta$-TCR$^+$ stage. This approach can facilitate studies aimed to assess the effects of environmental and genetic manipulations at various stages of T-cell development.

2 Materials

2.1 Cells

1. ESCs. Preferably lines that have shown lymphopoiesis in vitro, e.g.:
   (a) ESC lines R1, D3, E14K derived from 129/Sv mice.
   (b) ESCs from BALB/c, C57BL/6, and [C57BL/6x129]F2 mice.
2. Embryonic fibroblasts (EF).
   (a) Primary mouse embryonic fibroblasts [22].
3. OP-9 bone marrow stromal cells (OP-9 cells).
   (a) Obtained from the RIKEN Cell Repository (ID Number: RCB1124) (http://www.brc.riken.jp/lab/cell/english/index.shtml).
4. OP9-DL1 stromal cells (OP9-DL1).
   (a) OP-9 cells retrovirally transduced to express Delta-like 1 [19].

2.2 Reagents

1. Fetal Bovine Serum (FBS).

   Gibco: Cat. #12483-020.

   Gibco: Cat. #16141-079 (ES-Qualified) (*see* **Note 1**).

2. High Glucose Dulbecco's Modified Eagle's Medium (D-MEM).
   Sigma: Cat. # D-5671.
3. Alpha-Modified Eagle's Medium (alpha-MEM).
   Gibco: Cat. #12561-056.
4. 1× Phosphate Buffered Saline (PBS).
   Gibco: Cat. #14190-144.
5. 4-(2-Hydroxyethyl)-1-piperazineethanesulfonic acid (HEPES).
   Gibco: Cat. #15630-08.
6. Sodium Pyruvate.
   Gibco: Cat. #11360-070.
7. Gentamicin.
   Gibco Cat. #15750-060.
8. Penicillin/Streptomycin (P/S).
   Hyclone Cat. #SV30010.
9. Glutamax™.
   Gibco: Cat. #35050-061.
10. 2-Mercaptoethanol (2-ME).
    Gibco: Cat. #21985-023.
11. 2.5 % Trypsin.
    Gibco: Cat. #15090-046.
12. Mouse Leukemia Inhibitory Factor (mLIF).
    Millipore: Cat. #ESG-1107.
13. Mouse IL-7 (mIL-7).
    R&D: Cat. #407-ML.
14. Human Flt-3L (hFlt-3L).
    R&D: Cat. #308-FK.
15. Dimethyl Sulfoxide (DMSO).
    Sigma: Cat. #D2650.
16. Tissue culture-treated tissue culture plasticware.
    Any supplier.

2.3 Media/Reagent Formulations

1. ES media.
   (a) 500 mL of D-MEM.
   (b) 15 % ES-qualified FBS.
   (c) 1 M of HEPES.

   (d) 100 mM of Sodium pyruvate.
   (e) 50 mg/mL of Gentamicin.
   (f) 10,000 U/mL of P/S.
   (g) 200 mM of Glutamax™.
   (h) 55 mM of 2-ME.
2. OP9 media.
   (a) 500 mL of Alpha-MEM.
   (b) 20 % FBS.
   (c) 10,000 U/mL of P/S.
   (d) 50 mg/mL of Gentamicin.
3. Freezing media.
   (a) 90 % FBS v/v.
   (b) 10 % DMXO v/v.
4. Cytokines.
   (a) mLIF: Use at 10 U/μL.
   (b) mIL-7: Use at 1 μg/μL.
   (c) hFlt-3L: Use at 5 μg/μL.

3 Methods

The following protocol outlines the procedures for maintaining and differentiating ESCs on EFs or OP9/OP9-DL1 stromal cells, respectively. All incubations are performed in a humidified incubator at 37 °C with 5 % CO_2. All plasticware used are tissue culture treated for adherent cells. All centrifugation steps were performed at 300 ×*g* for 5 min at 4 °C.

3.1 Maintenance of ESC on EF Cells

ESCs are maintained on mitotically inactivated EF cells. EF cells are inactivated by irradiation at 30 Gy.

1. To prepare the EFs thaw first into 10 cm tissue culture plates. ESC media should be used for these cultures. Once the cells become confluent (after 2–3 days) they can be passaged with 0.25 % trypsin diluted with PBS.
2. To passage EFs first aspirate off the media on the plate and wash with 8 mL of PBS. Once PBS is aspirated, then add 4 mL of 0.25 % trypsin and incubate for 6 min. Prepare a 50 mL conical tube with 10 mL ESC media for cell collection. After the incubation, cells should become loosely adherent and can be easily washed off by rigorous pipetting with the addition of 4 mL of PBS. Collect the cell mixture in the 50 mL tube that

was prepared. Pellet the cells by centrifugation and resuspend in ESC media for replating at the appropriate split ratio. Note that each confluent 10 cm plate of EFs can be irradiated and plated into four 6 cm plates.

3. To seed ESCs to EF cells first thaw a vial of ESCs and seed to a 6 cm plate of confluent, irradiated EFs as described above. This coculture is grown in ESC media supplemented with mLIF. Seed no more than 3×10^4 cells to avoid contact between ESC colonies (*see* **Note 2**).
4. To passage ESCs, refer to the trypsin-mediated procedure described in **step 2** and use 2 mL of 0.25 % trypsin per 6 cm plate and wash with the addition of 2 mL of PBS. Once the cells are pelleted, resuspend in 3 mL ESC media with mLIF and filter through a 40 μm nylon mesh to remove large cell aggregates. Seed this cell suspension into a newly irradiated, 80 % confluent 6 cm plate of EFs. ESCs must be well dispersed in the plate to prevent large localized aggregates of ESCs from forming. Media supplemented with mLIF should be changed daily and cocultures split/passaged every 2 days to keep the ESC density below 80 % confluent (*see* **Note 2**).
5. To make frozen stocks of ESCs wash the plate with PBS and remove the cells as described in **step 2**. The cell mixture can be resuspended in 2 mL of freezing media to be aliquoted into freezing vials. Approximately 2–4 vials can be produced from a confluent 6 cm plate of ESC/EFs. Vials are frozen at −80 °C overnight and transferred to liquid nitrogen the next day (*see* **Note 3**).

3.2 Maintenance of OP-9 and OP9-DL1 Cells

1. To initiate an OP-9 culture first thaw a vial of OP-9 cells and plate to a 10 cm tissue culture plate in OP-9 media, until the cells reach 60 % confluency (2–3 days).
2. To passage OP-9 cells first aspirate media from the plate. Cells are removed as described in Subheading 3.1, **step 2**. The cell pellet can be resuspended in OP-9 media to be replated at the appropriate split ratio, not more sparsely than 1:8, typically at 1:4 ratio. OP-9s should be well dispersed to the plate to avoid localized cell aggregates, which can form adipocyte clusters that do not support ESC differentiation. OP9-DL1 cells are passaged in an identical manner as OP-9 cells (*see* **Note 4**).

3.3 ESC to T-Cell Differentiation on OP-9 and OP9-DL1 Cocultures

3.3.1 Day −3 to −2

1. Prepare ESC/EF cocultures 3 days before the ESC/OP9 coculture. Maintain the ESC/EF coculture as described above; ESC should be 80 % confluent for day 0.
2. Prepare 10 cm plates of OP-9s by thawing and expanding from day −3. Passage and maintenance procedures for OP-9 cells are described above. There should be at least two 80 %

confluent 10 cm plates of OP-9 cells ready for day 0: one to be seeded with ESCs and one to be passaged for OP-9 cell expansion.

3.3.2 Day 0

1. ESCs are removed with 0.25 % trypsin similar to the method described in Subheading 3.1, **step 4**. After the 6-min incubation cells are moved into 5 mL of PBS and no media is added. This cell mixture is centrifugated at $300 \times g$ for 5 min at 4 °C to isolate the cell fraction and also to induce low levels of auto-aggregation to remove adherent EF cells (*see* **Note 5**).
2. Cells are passed through a 40 μm nylon mesh and then resuspended in OP9 media for counting.
3. Seed 5×10^4 ESCs per 10 cm plate of OP9 cells in a total of 10 mL of OP9 media (*see* **Note 6**).

3.3.3 Day 3

1. Replace the media with 10 mL of fresh OP9 media. The old media should be centrifugated and any cells in the pellet should be recovered with fresh media to ensure that no cells are lost.

3.3.4 Day 5

1. Greater than 50 % colonies should have mesoderm-like morphology [22]. Cobblestone morphology should also be visible at this time point. Remove media without disturbing the coculture and centrifugate to pellet any cells to place back in coculture. The coculture should be passaged as described in Subheading 3.3.2, **steps 1** and **2**. Once the cells are counted, seed 6×10^5 cells per 10 cm plates of 80 % confluent OP-9 cells (*see* **Note 7**).
2. Fresh coculture should be grown in 10 mL OP9 media supplemented with hFlt-3L.
3. OP9-DL1 cells should be prepared similar to OP-9 cells at this time. Refer to Subheading 3.3.1 for instructions.

3.3.5 Day 7

1. Place passaged OP9-DL1 cells into 6-well tissue culture plates. Each confluent 10 cm plate can be plated into two 6-well plates (i.e., 12 wells) to be ready for day 8.

3.3.6 Day 8

1. The hematopoietic progenitors are now either loosely or no longer adherent to the OP-9 cells [23]. Remove the media from each plate carefully and very gently wash the stromal layer to remove all weakly adherent cells without disturbing the integrity of the adherent cells and the stromal cell layer (*see* **Note 8**).
2. Filter all collected cell mixture through a 40 μm nylon mesh to remove any stromal fragments, or undifferentiated ESC

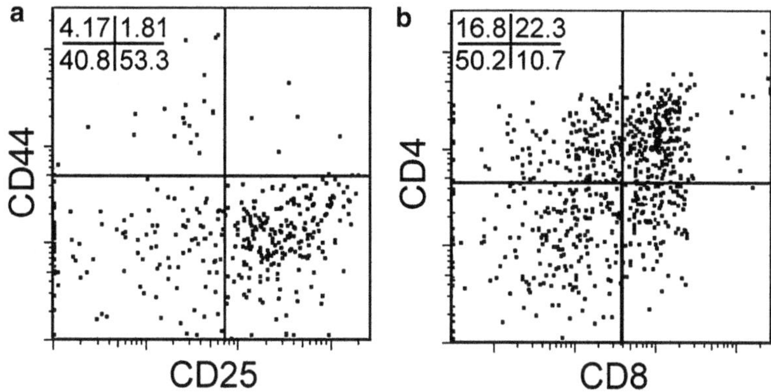

Fig. 1 T-cell development from an ESC/OP9-DL1 coculture. (**a**) On day 14, CD4 CD8 double-negative populations are characterized by their surface expression of CD44 and CD25. (**b**) CD4 CD8 double-positive cells, defined by their CD4 and CD8 co-expression, are observed on day 16. This population will make up the majority cells in the coculture at later time points

colonies, collected by accident. Centrifugate the collected cell suspension and resuspend in 3 mL of OP9 media, supplemented with mIL-7 and hFlt-3L, for each portion collected from a 10 cm plate. The non-adherent cells are plated into 6-well plates with OP9-DL1 cells. Each portion collected from a 10 cm plate is plated into 1 well of a 6-well plate.

3.3.7 Day 10

1. Change media by collecting, and centrifugating, all the media from each well, similar to Subheading 3.3, **step 3**. The cells are resuspended in 3 mL of OP9 media with added mIL-7 and hFlt-3L.

3.3.8 Day 12

1. The coculture is passaged by strong and rigorous pipetting to remove and disrupt the stromal cell integrity. The mixture of stromal cell fragments and non-adherent cells is filtered through a 40 μm nylon mesh to collect the non-adherent cells. Cells are pelleted by centrifugation, resuspended in 3 mL of OP9 media with mIL-7 and hFlt-3L, and plated onto fresh 6-well plates with 60 % confluent OP9-DL1 cells (*see* **Note 9**).

3.3.9 Beyond Day 12

1. Repeat **step 1** in Subheadings 3.3.7 and 3.3.8 by changing media every 2 days and stroma every 4 days. When coculture becomes overconfluent it is necessary to scale up to 10 cm plates (*see* **Note 10**). Figure 1 shows the effective generation of T-lineage cells from ESC-OP9-DL1 cocultures.

4 Notes

1. Prescreened lots of FBS for the propagation of undifferentiated ESCs are commercially available, but it is recommended to prescreen lots of FBS for ESC/OP9-DL1 cocultures. Different lots can be tested in parallel cocultures where ESCs are differentiated for 15–20 days. The quality of FBS is determined by the efficiency in generating T-lineage cells by their surface expression of T-cell markers, e.g., CD25, CD44, CD4, and CD8—*see* Fig. 1.
2. ESC colonies may become differentiated even if it is kept at below 80 % confluence. Localized cell aggregates will promote differentiation in clusters that must be removed by either physically picking them off of the plate or removing the cells with 0.25 % trypsin for 7 min instead of 5 min. Differentiated ESC colonies will appear less optically refractive and exhibit more adherent cell-like morphology with defined edges. Also by filtering the large cell aggregates, these cells can be removed.
3. Density of frozen stocks is important for subsequent thawing of cells. Due to their fast expansion rate it is advised to thaw 25 % of a 6 cm plate into a full 6 cm plate of EFs to allow space for ESCs to grow.
4. Each 1:4 split of OP9 or OP9-DL1 cells will take about 2 days to become 60–80 % confluent. Due to the potential heterogeneity of OP9 cells, splitting ratio beyond 1:8 may render the stroma functionally inert and unable to support T-cell development.
5. EF cells will aggregate faster than ESCs. After the centrifugation aggregation can be observed and rigorous pipetting is required to liberate any ESCs trapped in the cell aggregate. The resulting smaller cell aggregates can be excluded from the sample by passing it through a 40 μm nylon mesh filter.
6. Up to 3×10^5 cells can be seeded, depending on the duration of the experiment. For shorter experiments more cells can be used to compensate for the slow proliferation early on.
7. Up to 1×10^6 cells can be used per 10 cm plate, depending on the duration of the experiment. Micrographs of culture morphology can be found in ref. [24].
8. Cells with hematopoietic potential are in the weakly adherent fraction. Any adherent cells that are carried over into the next coculture step may take over the culture and negatively impact hematopoietic development.
9. Optimal T-cell development can be achieved by avoiding any disruption to the cell–cell contacts between hematopoietic and stromal cells. Nevertheless, it is necessary to change stroma every 4–5 days in order to provide sufficient support

for the expanding and differentiating cells. By pipetting rigorously loosely adherent cells can be detached from stromal cells. By breaking up the stromal layer into intermediate sized fragments their carryover can be limited.

10. When non-adherent round hematopoietic cells are observed to cover 80 % of the stromal layer at the bottom of the plate it is necessary to expand the coculture into 10 cm plates or more wells. When T-cell development reaches the end of the DN3 stage it will expand rapidly and extra stromal cells should be prepared to transfer the proliferating T-lineage cells. The timing of when each stage of T-cell development is reached varies and may depend on the type of ESC used. Therefore, flow cytometric analysis of the cultures at different time points, corresponding to cell passages, is recommended and highly informative [24].

References

1. Robertson EJ (1987) Teratocarcinomas and embryonic stem cells: a practical approach. IRL, Oxford, UK
2. Chen U, Kosco M, Staerz U (1992) Establishment and characterization of lymphoid and myeloid mixed-cell populations from mouse late embryoid bodies "embryonic-stem-cell fetuses". Proc Natl Acad Sci USA89:2541–2545
3. Gutierrez-Ramos JC, Palacios R (1992) In vitro differentiation of embryonic stem cells into lymphocyte precursors able to generate T and B lymphocytes in vivo. Proc Natl Acad Sci USA89:9171–9175
4. Potocnik AJ, Nielsen PJ, Eichmann K (1994) In vitro generation of lymphoid precursors from embryonic stem cells. EMBO J 13:5274–5283
5. Burt RK, Verda L, Kim DA, Oyama Y, Luo K, Link C (2004) Embryonic stem cells as an alternate marrow donor source: engraftment without graft-versus-host disease. J Exp Med 199:895–904
6. Nakano T (1995) Lymphohematopoietic development from embryonic stem cells in vitro. Semin Immunol 7:197–203
7. Nakano T, Kodama H, Honjo T (1994) Generation of lymphohematopoietic cells from embryonic stem cells in culture. Science 265:1098–1101
8. Yoshida H, Hayashi S, Kunisada T et al (1990) The murine mutation osteopetrosis is in the coding region of the macrophage colony stimulating factor gene. Nature 345:442–444
9. Cho SK, Webber TD, Carlyle JR et al (1999) Functional characterization of B lymphocytes generated in vitro from embryonic stem cells. Proc Natl Acad Sci USA96:9797–9802
10. Hirayama F, Lyman SD, Clark SC, Ogawa M (1995) The flt3 ligand supports proliferation of lymphohematopoietic progenitors and early B-lymphoid progenitors. Blood 85: 1762–1768
11. Hudak S, Hunte B, Culpepper J et al (1995) FLT3/FLK2 ligand promotes the growth of murine stem cells and the expansion of colony-forming cells and spleen colony-forming units. Blood 85:2747–2755
12. Hunte BE, Hudak S, Campbell D, Xu Y, Rennick D (1996) flk2/flt3 ligand is a potent cofactor for the growth of primitive B cell progenitors. J Immunol 156:489–496
13. Jacobsen SE, Okkenhaug C, Myklebust J, Veiby OP, Lyman SD (1995) The FLT3 ligand potently and directly stimulates the growth and expansion of primitive murine bone marrow progenitor cells in vitro: synergistic interactions with interleukin (IL) 11, IL-12, and other hematopoietic growth factors. J Exp Med 181:1357–1363
14. Lyman SD, Jacobsen SE (1998) c-kit ligand and Flt3 Ligand: stem/progenitor cell factors with overlapping yet distinct activities. Blood 91:1101–1134
15. Veiby OP, Lyman SD, Jacobsen SEW (1996) Combined signaling through interleukin-7 receptors and flt3 but not c-kit potently and selectively promotes B-cell commitment and differentiation from uncommitted murine bone marrow progenitor cells. Blood 88: 1256–1265
16. Pui JC, Allman D, Xu L et al (1999) Notch1 expression in early lymphopoiesis influences B versus T lineage determination. Immunity 11:299–308
17. Radtke F, Wilson A, Stark G et al (1999) Deficient T cell fate specification in mice with an induced inactivation of Notch1. Immunity 10:547–558

18. de Pooter RF, Cho SK, Carlyle JR, Zúñiga-Pflücker JC (2003) In vitro generation of T lymphocytes from embryonic stem cell-derived prehematopoietic progenitors. Blood 102: 1649–1653
19. Schmitt TM, Zúñiga-Pflücker JC (2002) Induction of T cell development from hematopoietic progenitor cells by delta-like-1 in vitro. Immunity 17:749–756
20. Schmitt TM, de Pooter RF, Gronski MA et al (2004) Induction of T cell development and establishment of T cell competence from embryonic stem cells differentiated in vitro. Nat Immunol 5:410–417
21. Zúñiga-Pflücker JC (2004) T-cell development made simple. Nat Rev Immunol 4(1):67–72
22. Robertson EJ (1997) Derivation and maintenance of embryonic stem cell cultures. Methods Mol Biol 75:173–184
23. de Pooter RF, Cho SK, Zúñiga-Pflücker JC (2005) In vitro generation of lymphocytes from embryonic stem cells. Methods Mol Biol 290:135–147
24. Holmes R, Zúñiga-Pflücker JC (2009) The OP9-DL1 system: generation of T-lymphocytes from embryonic or hematopoietic stem cells in vitro. Cold Spring Harb Protoc Pdb.prot5156 DOI:10.1101/pdb.prot5156.

Chapter 10

Development of Hematopoietic Stem and Progenitor Cells from Mouse Embryonic Stem Cells, In Vitro, Supported by Ectopic Human HOXB4 Expression

Sandra Pilat, Sebastian Carotta, and Hannes Klump

Abstract

Differentiation of pluripotent embryonic stem (ES) cells can recapitulate many aspects of hematopoiesis, in vitro, and can even generate cells capable of long-term multilineage repopulation after transplantation into recipient mice, when the homeodomain transcription factor HOXB4 is ectopically expressed. Thus, the ES-cell differentiation system is of great value for a detailed understanding of the process of blood formation. Furthermore, it is also promising for future application in hematopoietic cell and gene therapy. Since the arrival of techniques which allow the reprogramming of somatic cells back to an ES cell-like state, the generation of hematopoietic stem cells from patient-specific so-called induced pluripotent stem cells shows great promise for future therapeutic applications. In this chapter, we describe how to cultivate a certain feeder cell-independent mouse embryonic stem cell line, to manipulate these cells by retroviral gene transfer to ectopically express HOXB4, to differentiate these ES cells via embryoid body formation, and to selectively expand the arising, HOXB4-expressing hematopoietic stem and progenitor cells.

Key words Pluripotent stem cells, Embryonic stem cells, ES cells, Embryoid body, Hematopoietic stem cells, HSC, HSPC, HOXB4, Homeobox, Transcription factor, Stem cell expansion, Differentiation, Myeloid progenitors, Mass culture, Cell clones, Clonal expansion

1 Introduction

Hematopoietic stem cells (HSCs) are the central organizers of the hierarchical process of blood cell formation (hematopoiesis) in all adult vertebrates. Because of their ability to completely reconstitute multilineage hematopoiesis for a lifetime after transplantation, HSCs are a prime therapeutic target, for example in the context of treating genetic diseases of the hematopoietic system. In hematopoietic gene therapy, use of single HSC clones which have been

Nicholas Zavazava (ed.), *Embryonic Stem Cell Immunobiology: Methods and Protocols*, Methods in Molecular Biology, vol. 1029, DOI 10.1007/978-1-62703-478-4_10, © Springer Science+Business Media New York 2013

molecularly characterized and expanded to therapeutically effective amounts prior to transplantation is highly desirable. However, selective in vitro expansion of HSCs has remained a challenging task, despite improvement of growth conditions allowing better maintenance and even some modest expansion of HSCs, in vitro [1–3]. In contrast, ectopic expression of certain transcription factors, for example of the homeodomain transcription factor HOXB4, has proven more efficient in mediating expansion of hematopoietic stem and progenitor cells (HSPCs) of humans and mice, in vitro and in vivo [4–6]. In fact, human *HOXB4* was the first gene shown to mediate a profound HSPC expansion in vitro and in vivo when ectopically expressed in murine bone marrow cells without inducing leukemia [5]. Another major hurdle is that the individual identity of HSCs has remained obscure, although significant progress has been made in defining surface markers associated with HSCs, thus allowing substantial enrichment of long-term repopulating activity from bone marrow [7–10]. Hence, isolation and direct manipulation of HSC clones have remained technically impossible, so far. At present, the only cell types allowing for application of gene targeting techniques, in vitro selection, and expansion of defined clones with high efficiency are pluripotent embryonic stem (ES) and induced pluripotent stem (iPS) cells. Therefore, pluripotent stem cells are an attractive alternative as a source for generating tailored HSCs, in vitro. After in vitro differentiation of ES cells, some progeny cells are capable of mediating long-term hematopoietic repopulation after transplantation into recipient mice, in vivo, when HOXB4 is ectopically expressed above a certain threshold [11–15] and can induce transplantation tolerance of allografts [16]. How HOXB4 mechanistically enhances this conversion towards HSCs is currently not understood. So far, expression profiling in combination with functional analyses suggested that this transcription factor alters the same common pathways known to be important for controlling stem cell self-renewal and differentiation, both in adult as well as in ES cell-derived hematopoietic stem and progenitor cells [17]. In this chapter, a basic protocol is presented for cultivation (Subheading 3.1) and differentiation of a mouse embryonic stem cell line by "embryoid body" (EB) formation in semisolid medium (Subheading 3.3). Because retroviral transduction of the human homeodomain transcription factor HOXB4 into undifferentiated ES cells enables subsequent outgrowth of HSPCs after EB-differentiation, in vitro, without any stroma cell support [12, 17], production of a retroviral HOXB4 expression vector, titer determination, and retroviral gene transfer (transduction) into undifferentiated ES cells are explained in Subheading 3.2. Pluripotent stem cells could significantly promote regenerative medicine by allowing the usage of techniques not yet established

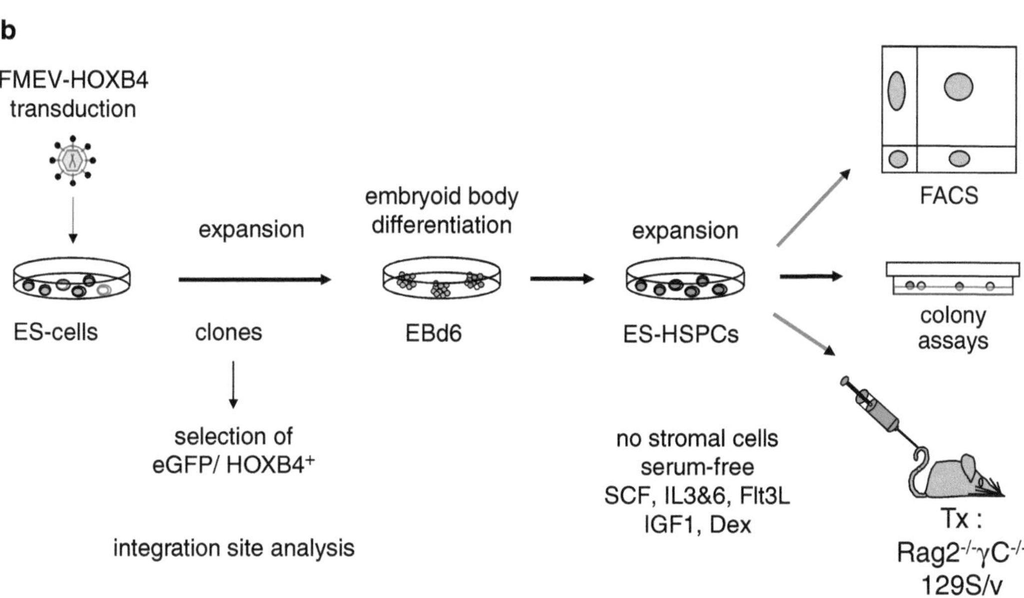

Fig. 1 (**a**) Schematic representation of the retroviral coexpression vector for HOXB4. Transcription is driven by the U3 enhancer/promoter within the long terminal repeat (LTR) of the retroviral vector. During translation, the 2A-esterase of foot-and-mouth disease virus (FMDV) inhibits peptidyl-transferase activity in the ribosome [31], consequently leading to separated eGFP and HOXB4 in constant molar ratios and in a non-proteolytical fashion. Abbreviations: *LTR* long terminal repeat, *U3* unique region 3′, *R* repeat, *U5* unique region 5′, *SD* splice donor, *SA* splice acceptor within the 5′ untranslated region, ψ *wPRE* woodchuck hepatitis posttranscriptional regulatory element. (**b**) Flowchart of retroviral transduction of ES cells, cultivation of HOXB4-expressing clones, embryoid body (EB) differentiation, expansion of ES cell-derived hematopoietic stem and progenitor cells (HSPCs) and readout systems, in vitro (FACS and colony formation assays) and in vivo (mouse transplantation; not explained in this chapter)

for somatic stem cells. The recent finding that somatic cells can be reprogrammed back to a pluripotent state [18–20] has rapidly enabled the generation of patient-specific iPS cells, in vitro [21], and thus will open avenues for the future generation of patient-specific HSCs, in vitro. The subsequent protocol is intended to convey you the ability to propagate mouse ES cells (CCE line) (Subheading 3.1), differentiate them via embryoid bodies (Subheading 3.3), and expand ES cell-derived hematopoietic cells (Subheading 3.4) after HOXB4 transduction (Subheading 3.2). An outline is depicted in Fig. 1b.

2 Materials

2.1 Mouse Embryonic Stem (CCE Line) Cell Culture

1. "Knock-out" Dulbecco's Modified Eagle Medium (K.O. DMEM) (Gibco/BRL, Bethesda, MD, Cat. No.#10829-018) supplemented with 15 % (v/v) fetal calf serum (FCS) (*see* **Note 1**), 1 % (v/v) Penicillin/Streptomycin (PenStrep) (Sigma-Aldrich, Cat. No. #P0781), 2 mM L-Glutamine (Sigma-Aldrich, Cat. No. #G7513), 1.5×10^{-4} M Monothioglycerol (MTG) (Sigma-Aldrich, Cat. No. #M6145) (*see* **Note 2**) in 1× PBS, and 10 ng/ml Leukemia Inhibitory Factor (LIF, R&D #449-L).
2. 0.1 % (w/v) gelatin (Sigma, Cat. No. #G1890) dissolved in phosphate-buffered saline (PBS) and autoclaved.
3. Solution of Trypsin (0.25 %) from Gibco/BRL.
4. 25 cm^2 cell culture flasks for adherent cells from Nunc (Cat. No. #136196).

2.2 Methylcellulose Stock Solution

Material: Methocel (Fluca, Cat. No.#64630), IMDM powder (Gibco/BRL, Cat. No. #12100-046), Monothioglycerol (MTG) (Sigma-Aldrich, Cat. No. #M6145), a sterile Erlenmeyer flask (with screw cap).

1. Put a stirring staff into the flask and weigh, rinse with sterile cell culture-quality water, then add 450 ml H_2O, and boil on a heater. In the meanwhile, prepare 20 g methylcellulose powder per liter solution. As soon as the water starts boiling, add methylcellulose, mix with magnetic stirrer, simmer for about 10 min, then boil up once, and subsequently let cool down to room temperature slowly.
2. For a 2× IMDM medium stock mix 3.025 g/l sodium-bicarbonate, IMDM powder for 1 l (*see* **Note 3**), 1.0 ml of an MTG 1,000× stock (=150 mM). Then add 450.0 ml sterile H_2O, dissolve, and filter sterile. Finally, add Penicillin and Streptomycin to a final concentration of 1% (v/v) from a commercial 100x PenStrep stock solution and 2mM L-glutamine.

As soon as the flask containing the viscous methylcellulose solution is hand-hot (approx. body temperature), move it into a sterile work bench containing a magnet-stirrer. Pour 2× IMDM medium to the methylcellulose with simultaneous stirring (the color of the indicator may change from red to yellow). Because the final weight must be 1,000 g, subtract the total weight of the mixture from the weight of the flask + stirrer and adjust with sterile cell culture-quality H_2O. Place the flask in a cold room (4 °C) and let cool down overnight with simultaneous slow stirring. During the first 2 h, swirl manually approximately every 30 min to achieve a homogenous solution. Next day pour 40 ml aliquots per 50 ml tubes under the sterile hood, shock-freeze in liquid N_2, and store at −80 °C. Prior to use after thawing, centrifuge at 720×*g* for 10 min to remove large fibrous cellulose pieces. Use only 30–35 ml of the supernatant for further steps.

2.3 Methylcellulose Differentiation Medium

Components:
IMDM (Gibco/BRL, Cat. No. #21980-032), pretested FCS, PenStrep (Sigma-Aldrich, Cat. No. #P0781), L-Glutamine (Sigma-Aldrich, Cat. No. #G7513), Human Transferrin (Sigma, Cat. No.# T0665), Protein Free Hybridoma Medium (PFHM-II) (Gibco/BRL, Cat. No. #12040-051), Monothioglycerol (MTG) (Sigma-Aldrich, Cat. No. #M6145), Ascorbic Acid (Sigma-Aldrich, Cat. No. #A4544).

For 260 ml of semisolid embryoid body differentiation media mix the components as follows:

IMDM	66.0 ml
FCS (f.c. 15 %)	39.0 ml
100× PenStrep solution	1.3 ml
L-Gln (f.c. 2 mM)	2.6 ml
Human transferrin (f.c. 300 μg/ml)	4.5 ml
PFHM-II (f.c. 5 %)	13.0 ml
150 mM MTG (f.c. 4×10^{-4} M)	0.7 ml
Wait for approx. 10 min, then add	
Ascorbic acid[a] (f.c. 50 μg/ml)	2.6 ml
Methylcellulose solution (f.c. 50 %)	130.0 ml
Total	260.0 ml

[a]For preparation of an ascorbic acid stock solution, dissolve 5 mg/ml in cold (4 °C) tissue culture-grade H_2O, filter sterile, make 1.5 ml aliquots, and store at −80 °C . After thawing, use only once and discard excess.

2.4 Culture Medium for ES Cell-Derived Hematopoietic Cells ("SCM")

StemPro34 + Nutrient Mix (Gibco/BRL #10639) containing mSCF (100 ng/ml), mIL-3 (2 ng/ml) (R&D 403-ML; stock: 2 μg/ml in PBS, 0.1 % BSA), mIL-6 (5 ng/ml) (R&D 406-ML; stock: 5 μg/ml in PBS, 0.1 % BSA), Flt3-L (10 μg/ml) (R&D 427-FL; stock: 10 μg/ml in PBS, 0.1 % BSA), IGF-1 (40 ng/ml) (Sigma I-1271; stock: 40 μg/ml IGF-1 in 3.4×10^{-4} M acetic acid, PBS, 0.1 % BSA), Dexamethasone (1 μM) (Sigma D4902; stock: 1 mM in ethanol).

Combine StemPro34 and nutrient mix, add dexamethasone (f.c. 1 μM) and IGF-1 (f.c. 40 ng/ml). Aliquot 40–45 ml in 50 ml tubes. Shock-freeze in liquid N_2 and store at −80 °C. Add cytokines immediately before use and store for no longer than 2 weeks at 4 °C.

2.5 Antibodies for Flow Cytometry

All the following anti-mouse antibodies were purchased from eBioscience, San Diego, CA, USA: GR-1 (clone RB6-8C5), CD11b (clone M1/70), CD41 (clone ebioMWReg30), CD117 (c-Kit; clone 2B8), CD31 (PECAM, clone 390), and CD45 (clone 30-F11).

2.6 Retrovirus Production

1. All cell lines were purchased from ATCC-LGC (http://www.lgcstandards-atcc.org/). HEK 293T/17 cells were used for production of virus (ATCC-Nr. CRL-11268), mouse SC1 embryonic fibroblasts for titration of ecotropic retroviruses (ATCC-Nr. CRL-1404).
2. All chemicals necessary for calcium phosphate transfection were purchased from Sigma-Aldrich. A detailed protocol for preparation of the required buffers can be usually found in common molecular biology textbooks [22].
3. Prepare a 4 mg/ml (1,000×) stock solution of Protamine sulfate (Sigma P4020) in cell culture-quality H_2O. Sterile-filter, aliquot, and store at −20 °C.
4. Prepare a 25 mM (1,000×) stock solution of Chloroquine (Sigma C6628), in cell culture-quality H_2O. Sterile-filter, aliquot, and store at −20 °C.
5. Sterile 33 mm Milex-GP filter units, 0.22 μm (Millipore, Cat. Nr. SLGP033RS).
6. Expression plasmids for ecotropic envelope and MLVgag-pol can be purchased online via Addgene (www.addgene.org). ecotropic MLVenv (pHCMV-EcoEnv; plasmid nr. 15802); MLVgag/pol (MSCV-gag/pol; plasmid nr. 14887).

3 Methods

3.1 Mouse Embryonic Stem Cell Culture

For hematopoietic differentiation studies, we use clones of the CCE ES cell line [23] into which a retroviral expression vector for HOXB4 is introduced [12, 14, 16, 24, 25] (*see* **Note 4**). CCE cells can be cultured in LIF without embryonic fibroblast feeder cells. However, we try to avoid culturing the ES cells for more than 2–3 weeks (*see* **Note 5**).

1. CCE cells should be grown on gelatinized T25 flasks. Prepare the flasks by pipetting 0.1 % sterile gelatin in 1× PBS solution until the bottom is well covered and let stand at room temperature for at least 30 min. Remove the PBS immediately before use.
2. Rapidly thaw vial(s) containing frozen ES cells in a 37 °C water bath exactly to the point where there still is a small remnant of ice inside the vial. Immediately transfer the cell suspension from the cryotube into a tube containing at least ten times as much prewarmed ES-cell medium (Subheading 2.1) and pellet the cells by centrifugation at 260 × *g* for 10 min.

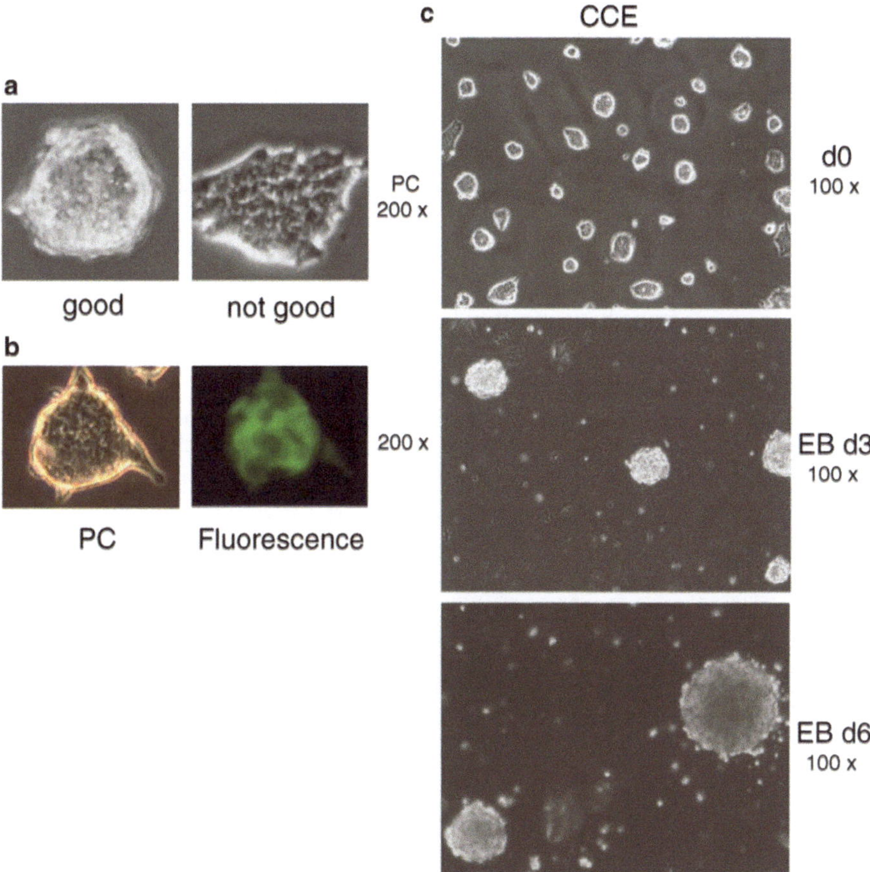

Fig. 2 (**a**) Morphology of ES-cell colonies (CCE cell line). Undifferentiated colonies should have a smooth edge (the so-called shiny border appearance); boundaries of single cells should not be visible (compare *left* "good" with *right* "not good" colony). (**b**) An ES-cell colony is shown after transduction with a retroviral GFP expression vector (FMEV-eGFP+wPRE), in phase contrast (PC) and fluorescence (*right picture*). The inhomogeneous fluorescence may be due to silencing of the vector in some cells of the colony. (**c**) The development of embryoid bodies (EBs) on day 0 (d0), day 3 (EB d3), and day 6 (EB d6) is depicted. The *spherical shape* of the cystic EBs can be clearly recognized on the EB d6 picture

3. Remove the supernatant with a pipette, carefully resuspend the cells in fresh ES-cell medium, transfer into the gelatin-coated flasks, and culture them in an incubator containing 5 % CO_2 and a H_2O-saturated atmosphere at 37 °C (*see* **Note 6**) (Fig. 2a).
4. For passaging ES cells, remove the medium, wash the cells once with prewarmed 1× PBS, and add 2 ml of trypsin–EDTA solution prewarmed to 37 °C. Immediately remove the trypsin solution again so that the colonies only stay "wet." Incubate at 37 °C for about 3 min, until cell–cell boundaries become visible which is then the time to shake the flask for colony dissociation (*see* **Note 7**).
5. Dilute about 1:5–1:10 in fresh ES-cell medium and incubate again.

3.2 Retroviral Transduction of Embryonic Stem Cells

For the present protocol, HOXB4 was introduced into ES cells by retroviral transduction prior to differentiation, thus allowing for extensive molecular characterization of purified ES-cell clones prior to further use. Although retroviral vectors are a very efficient means for transfer into and stable expression of any cDNA in most cells, they are often silenced in pluripotent cells. As a consequence, long-term expressing ES cells have to be isolated by repeated purification, for example by flow cytometry-based sorting. Another possibility to avoid extensive silencing is to perform retroviral transduction after ES-cell differentiation and dissociation of EBs. However, such polyclonal ES cell-derived hematopoietic cultures containing undefined retroviral integrations may show effects only due to insertional mutagenesis and unrelated to HOXB4 expression. For this work, we coexpressed eGFP and HOXB4 via the 2A-esterase of the picornain foot-and-mouth disease virus (FMDV), which separates both proteins cotranslationally (Fig. 1a). In contrast to IRES-elements, cotranslational separation leads to a stable expression of both proteins at a constant molar ratio [6, 26, 27], enabling a much better estimation of the amounts of HOXB4 by measuring GFP fluorescence [27]. Expression was driven by an FMEV-based retroviral vector backbone, which we have shown to mediate long-term high expression in embryonic and adult stem cells [6, 12]. For virus production, HEK 293T/17-cells were transfected by the calcium phosphate precipitation method, described in detail by Kingston et al. [22].

3.2.1 Production of Ecotropic Retrovirus-Containing Cell Culture Supernatants

1. In the late afternoon or evening, seed 5×10^6 293T/17-cells in a 9 cm tissue culture dish in 10 ml DMEM supplemented with 10 % FCS and 2 mM L-glutamine.
2. Next day, mix expression plasmids encoding ecotropic envelope (5 μg), MLV gag-pol (10 μg), and 5 μg of the FMEV-FMEV-eGFP2AHOXB4+Pre vector (FMEV-eGFP+Pre as a control) in 438 μl water and add 50 μl 2.5 M $CaCl_2$.
3. Add the DNA/$CaCl_2$ solution to 500 μl 2× Hepes-buffered saline by bubbling air in a conical tube; vortex briefly.
4. Immediately before transfection, remove old medium and add fresh medium containing 25 μM chloroquine.
5. Add DNA-mix to cells and swirl plate gently.
6. After 6–12 h, exchange medium for 10 ml fresh medium without chloroquine.
7. Next morning, remove old medium and add 7 ml fresh medium containing 20 mM Hepes-buffer (pH 7.4). Harvest the first virus-containing supernatant in the evening, which is about 30 h post transfection. To do so, take up the supernatant with a 10 ml syringe and pass it through a Milex-GP (Millipore) sterile filter (0.22 μm). It is recommendable to

collect 1 ml of the supernatant separately in a 1.5 ml sterile reaction tube for later titration. Store supernatants at –80 °C.

8. Collect viral supernatants every 12 h up to 72–84 h post transfection and store at –80 °C until further use.

3.2.2 Titration of Retroviral Supernatants

For titration of ecotropic virus supernatants we routinely use mouse SC1 fibroblasts growing in 24-well cell culture plates.

1. Prepare plates by seeding 1×10^5 exponentially growing SC1 cells per 24-well in 1 ml DMEM containing 10 % FCS, 4–6 h prior to virus transduction. We recommend to test at least three different dilutions of the virus supernatant, e.g., 10, 25, and 50 μl supernatant. Seed an additional four wells as a negative control.
2. After 4–6 h, exchange the medium for 500 μl DMEM supplemented with 10 % FCS and 8 μg/ml protamine sulfate per well.
3. Prepare the virus supernatants for titration by adding 500 μl DMEM/10 % FCS to 10, 25, and 50 μl aliquots of each viral supernatant in a sterile 1.5 ml reaction tube. Keep on ice until use.
4. Add the 500 μl virus dilutions to the cells, seal the plates with parafilm, and centrifuge at $800 \times g$ for 60 min at 32 °C.
5. Remove parafilm sealing and place cells into the 37 °C/5 % CO_2 incubator.
6. Next day, remove the supernatant from cells and replace with 1 ml fresh DMEM/10 % FCS medium.
7. On day 3 post transduction, remove medium, wash cells once with prewarmed 1× PBS, trypsinize the cells, and resuspend in 1× PBS containing 4 % FCS. It may be necessary to remove larger cell aggregates by filtration through a gaze-filter prior to flow cytometrical analysis to avoid clogging of the FACS-nozzle.
8. Determine the percentage of GFP-positive cells by flow cytometry.
9. The apparent titer/ml (termed "GFP transducing units," GTUs) can be determined according to the following formula:

$$\frac{\text{proportion of GFP positive cells} \times \text{seeded cell numbers at the day of infection} \times 2}{\text{volume of virus supernatant (in ml)}}$$

3.2.3 Transduction of ES Cells

With the following protocol, we routinely achieve a transduction efficiency of about 10 % transgene positive cells 2 days after transduction.

1. Trypsinize and count ES cells. Seed 3×10^5 cells in ES-cell medium supplemented with 20 mM Hepes-buffer (pH 7.4)

and 4 µg/ml protamine sulfate to each well of a 12-well plate for suspension cultures. The volume should not exceed 2 ml; otherwise use a 6-well plate.

2. Add the virus-containing supernatant at a multiplicity of infection (MOI) of 10 (*see* **Note 8**) and centrifuge plates at 711 ×*g* for 60 min at 32 °C.
3. After centrifugation, collect the ES cells, add ES-cell medium to a final volume of 5 ml, and transfer the suspension into a gelatinized 25 cm² flasks (*see* Subheading 3.1). Incubate overnight at 37 °C/5 % CO_2 and exchange medium next morning.
4. 2–3 days after transduction, continue to propagate the colonies as described in Subheading 3.1. Determine the transduction efficiency with an aliquot of the single-cell suspension after trypsinization. For single-clone isolation, flow cytometrical sorting is most convenient (*see* **Note 9**).

3.3 ES-Cell Differentiation

1. 48 h before starting embryoid body differentiation, transfer ES cells to IMDM-ES medium (in which the DMEM of the ES-cell medium is simply exchanged for IMDM) in a gelatinized T25 flask (*see* **Note 10**).
2. After 2 days in IMDM-ES medium, harvest cells by trypsinization, count, and check for viability (*see* **Note 11**). Wash cells in prewarmed 1× PBS, resuspend in fresh IMDM-ES medium without LIF, and dilute to a concentration of about 1,500 cells/ml.
3. Suspend in methylcellulose differentiation medium, plate on 20 cm bacterial petri dishes, and incubate at 37 °C for the first 2 days.
4. On day 3, exchange about 50 % of the methylcellulose medium.
5. On day 5, completely exchange the medium. To do so, use a pipette with a wide opening, slightly tilt the petri dish, carefully rinse down the EBs, and collect them in a 50 ml Falcon tube. They will visibly precipitate. Pellet them with low g-force by centrifugation at 18 ×*g* for 2 min.
6. Discard the supernatant, carefully resuspend the pelleted EBs in fresh warm methylcellulose differentiation medium, and transfer the suspension onto a new 20 cm bacterial petri dish. Remove air bubbles and place back into the incubator. The appearance of EBs during the procedure is shown in Fig. 2c.
7. On day 6, collect the EBs again as on day 5, wash with prewarmed 1× PBS, and pipette 5–7 ml trypsin solution on top of the pelleted EBs. Do not resuspend using a pipette. Instead,

swirl carefully. Place the tube into a 37 °C water bath for about 2–4 min and agitate every 30 s.

8. As soon as the solution becomes cloudy, inactivate the trypsin by adding pure FCS (ca. 3 ml) and immediately vigorously resuspend with a 5 ml pipette (otherwise the cells will clump together). Filter the cell suspension through a 70 μm nylon filter (*see* **Note 12**).
9. Pellet cells at 260 × *g*/5 min, wash at least twice (*see* **Note 13**), resuspend in fresh 1× PBS, and determine the cell numbers and viability.
10. Dilute cells to 3.5×10^6/ml in hematopoietic serum-free medium (see below) and plate in a 9 cm suspension cell petri dish.

3.4 Outgrowth and Expansion of Hematopoietic Stem and Progenitor Cells

3.4.1 Cultivation of ES Cell-Derived Hematopoietic Cultures (ES-HSPCs)

ES-HSPC cultures must be taken care of daily.

1. Culture ES-HSPCs in SCM at a density of 3.5×10^6/ml during the first 2 days, and then at $2–2.5 \times 10^6$/ml for the rest of the culture time. Exchange about half of the medium for fresh medium every day.
2. Incubate the cells at an increased CO_2 partial pressure (8 %) and avoid longer drops of the CO_2 concentration by keeping manipulation steps outside of the incubator as short as possible. GFP-transduced control cells without ectopic HOXB4 expression will only continue to proliferate for about 2 weeks and then stop growing. Only the HOXB4 cultures will continue to grow exponentially under these conditions, at least for more than 200 days (Fig. 3a, b). Regularly determine the percentage of dead cells in culture. From time to time, dead and differentiated cells adhering to the petri dish have to be removed by Ficoll gradient centrifugation (*see* **Note 14**).

3.4.2 Removal of Dead and Differentiated Cells by Ficoll Gradient Centrifugation

1. Transfer the cell suspension into a 50 ml Falcon tube and carefully pipette cold (4 °C) Ficoll (density gradient separation medium, density 1.1 g/ml) under the cell suspension using a Pasteur pipette.
2. Centrifuge at 720 × *g* for 10 min (~2,000 rpm in a Heraeus Megafuge 10).
3. Take cell suspension including the interphase (dead cells and debris are at the bottom of the tube), place into new Falcon tube, and pellet at 260 × *g* for 5 min.
4. Remove supernatant and resuspend pelleted cells in SCM.
5. Check the quality of cell suspension (cell numbers and viability, *see* **Note 11**) and continue to culture at a density of $2–2.5 \times 10^6$/ml.

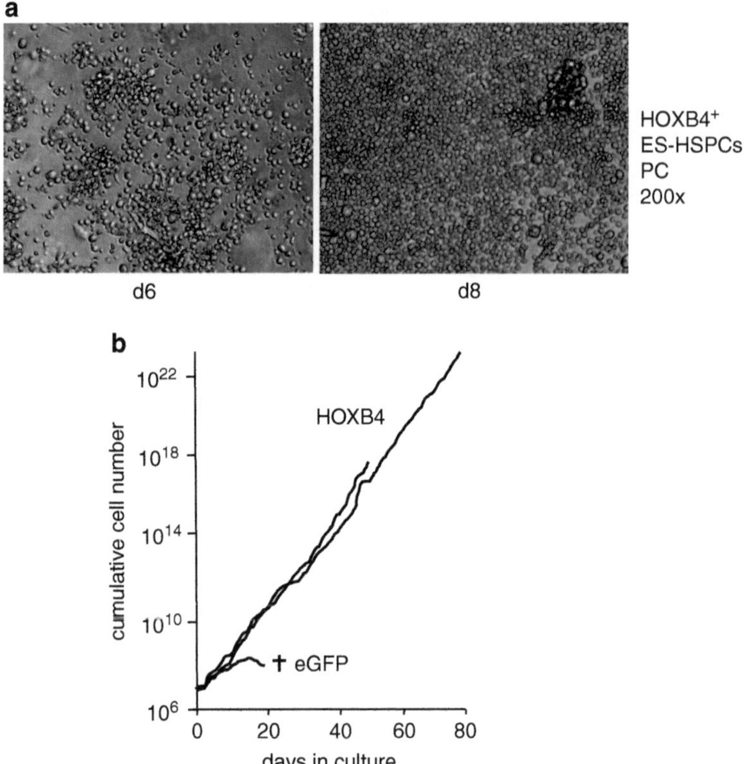

Fig. 3 (**a**) Hematopoietic suspension cultures ectopically expressing HOXB4 (HOXB4+ ES-HSPCs) at 6 and 8 days after dissociation of the day 6 EBs (d6 and d8, respectively). (**b**) Whereas HOXB4-transduced cultures continue to grow exponentially for >200 days (two separate clones are shown), control cultures without enforced HOXB4 expression usually cease growing and die after about 20 days under the conditions described in this protocol (termination of eGFP control vector-transduced cell growth is depicted with a cross)

3.4.3 Characterization of ES-HSPC Cultures by Flow Cytometry

Multicolor flow cytometry facilitates the analysis of ES-HSPCs and enables the detection and possible enrichment of the earliest definitive hematopoietic progenitors in culture over time by simultaneous detection of indicative surface markers on each cells. For example, CD41+ cells (gp IIa, Itga2b, αIIβ integrin) detected during ES-cell differentiation, in vitro, and in the embryo, in vivo, are restricted to a hematopoietic fate and define the onset of primitive and definitive hematopoiesis [28, 29]. Figure 4 shows simultaneous expression of CD41, CD45, and c-Kit on ES-HSPC cultures cultured for 21 days after EB dissociation.

1. For staining the ES-HSPCs with these antibodies, count the cells and use about 5×10^6 cells per staining reaction in a 1.5 ml reaction tube. Wash with 1× PBS at $260 \times g$ for 10 min.

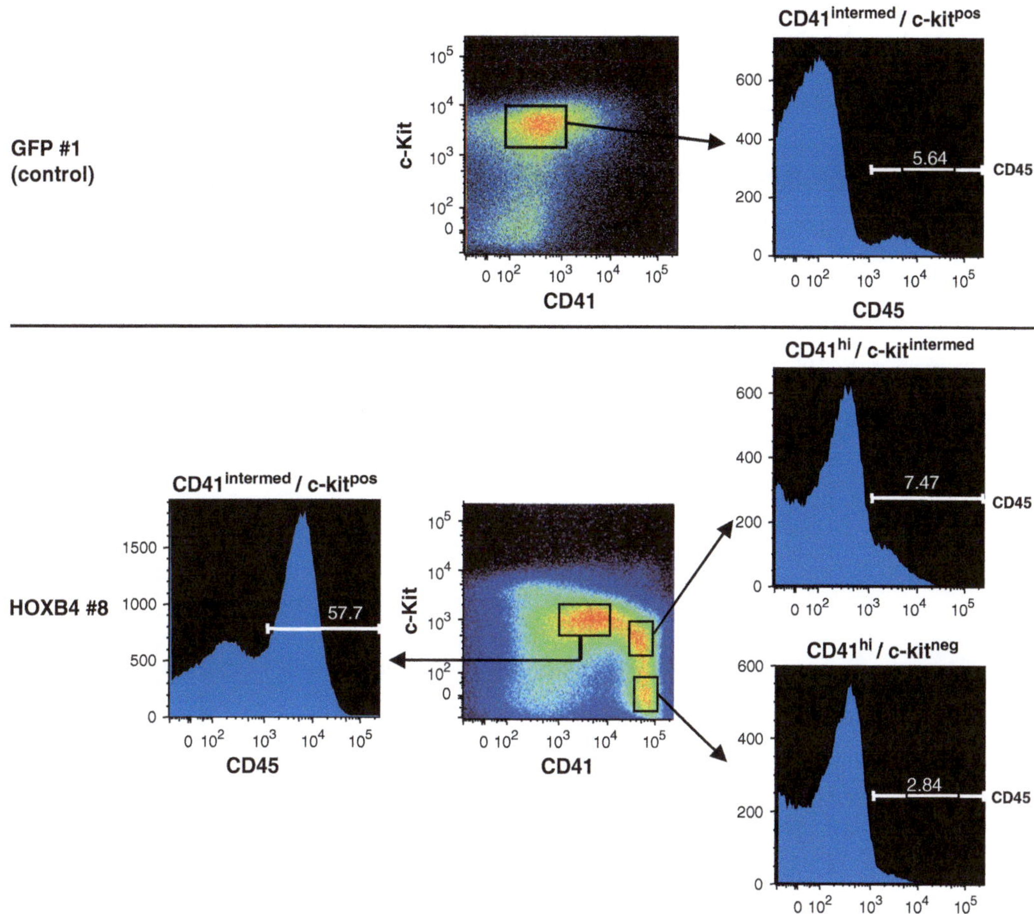

Fig. 4 HOXB4 expression leads to the selective expansion of CD41$^+$ hematopoietic progenitors. Flow cytometrical analysis is shown of day 10 ES-HSPC cultures derived from ES-cell clones either expressing the GFP control vector (*upper panel*, GFP #1) or HOXB4 (*lower panel*, HOXB4 #8). In contrast to control cultures, which lack cells expressing high amounts of CD41 and contain only few cells positive for CD45, the HOXB4 cultures contain a major amount of CD41^{hi} cells, which tend to be negative for CD45 expression, whereas CD41$^{lo/intermediate}$ cells which have upregulated c-Kit expression also become CD45$^+$. CD41^{hi} cells display blast-like morphology (*see* Fig. 5) and are capable of reconstituting the heterogeneity of the suspension culture when purified flow cytometrically and cultivated (data not shown)

2. Resuspend the cells in 100 μl PBS/4 % FCS containing 100–200 ng uncoupled anti-CD16/32 antibody and incubate for 30 min. This will block eventually present Fc-receptors, which may lead to a false-positive signal by binding the specific, fluorochrome-coupled antibodies.

3. Wash once with 1× PBS and resuspend the cells in 100 μl PBS/4 % FCS containing different combinations of the following antibodies according to the "fluorescence minus one" scheme depicted in Table 1 [30] (*see* **Note 15**).

Table 1
Staining reactions for simultaneous detection of GFP, CD41, CD45, and CD117 (c-Kit) on GFP+ or GFP/HOXB4+ ES-HSPCs via multicolor FACS, according to the "fluorescence minus one" method for correct compensation [30]

	Cells/reaction	GFP or GFP/HOX	CD41-PE	CD117-PerCP-Cy5.5	CD45-APC
FMO-control	1×10^5 cells	+	−	−	−
FMO-control	1×10^5 cells	+	+	−	−
FMO-control	1×10^5 cells	+	−	+	−
FMO-control	1×10^5 cells	+	−	−	+
Measurement	5×10^5 cells	+	+	+	+

4. After incubation for about 60 min at 4 °C in the dark, wash the cells twice, resuspend in 300 ml 1× PBS/4 % FCS, and analyze on a (at least) four-channel flow cytometer capable of detecting the employed fluorochromes (*see* **Note 16**).

3.4.4 Morphological Evaluation of ES-HSPC Cultures by Cytological Staining

This method complements FACS analysis and often provides a lot of additional important information on the quality and composition of the cultures. An example of HOXB4+ ES-HSPCs after flow cytometrical sorting is shown in Fig. 5.

1. Wash approximately 5×10^4 cells in 1× PBS/2 % FCS and dilute in 100 μl of 1× PBS/1 % BSA. Keep your samples on ice.
2. Place slides and filters into appropriate slots in the cytospin centrifuge with the cardboard filters facing the center of the cytospin. Pipette about 200 μl of cold 1× PBS/1 % BSA into each of the wells and spin at $150 \times g$ for 3 min.
3. Quickly place about 100 μl of each sample into the appropriate wells of the cytospin and spin at $150 \times g$ for 7 min.
4. Remove the filters from their slides without contacting the smears on the slides.
5. Dry the slides at room temperature.
6. Cover the slide with undiluted May-Grünwald staining solution for 5 min (*see* **Note 17**).
7. Rinse slide with distilled H_2O.
8. Stain with diluted Giemsa solution (according to the manufacturer) for 20 min.
9. Thoroughly rinse slide with distilled H_2O and let dry before microscopy.

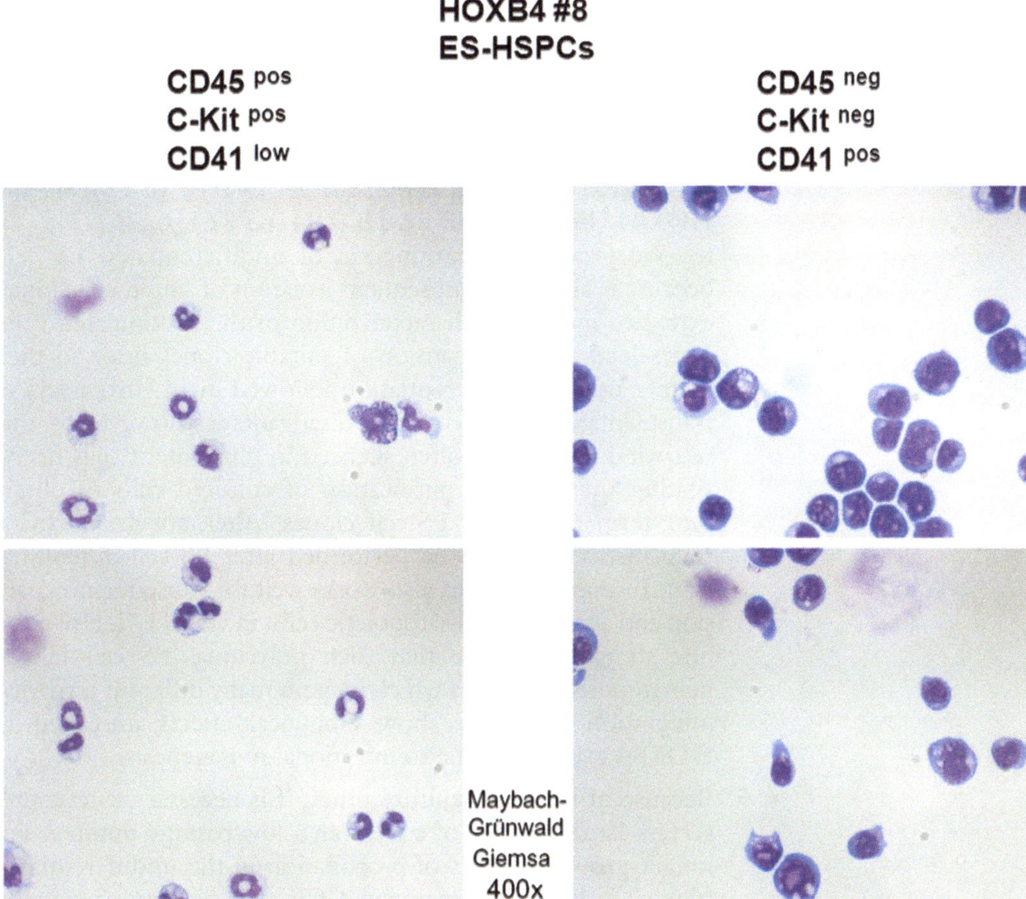

Fig. 5 ES cell-derived hematopoietic cells (ES-HSPCs) can develop towards mature granulocytes, entirely in vitro. EB-derivatives from an ES-cell clone ectopically expressing HOXB4 (HOXB4 #8) are shown. The ES-HSPCs, grown for 21 days in serum-free medium containing appropriate cytokines [12], were flow cytometrically sorted into $CD41^{lo}$/c-Kit^{+}/$CD45^{+}$ (*left panels*) and $CD41^{+}$/$ckit^{neg}$/$CD45^{neg}$ (*right panels*) subpopulations, spun onto glass slides and stained with May-Grünwald-Giemsa. $CD41^{lo}$/c-Kit^{+}/$CD45^{+}$ cells display a mature neutrophil phenotype. In contrast, cells expressing high amounts of CD41 show a more blast-like phenotype

4 Notes

1. It is highly recommendable to pretest FCS batches of diverse companies for their ability to support maintenance of ES cells in the undifferentiated state and reserve a larger amount of the positively identified batch. Many companies offer storage of the serum until call.
2. We usually prepare 100 ml of MTG solution in 1× PBS, and, after sterile filtration through 20 μm filters, store 1 ml aliquots at −80 °C.

3. The osmolarity of the final solution is crucial for these experiments. Thus, we do not recommend ready-to-use liquid IMDM in this case. The osmolarity can be adjusted more precisely when weighing out the powder.
4. ES cells are transduced with an FMEV-based retroviral expression vector containing eGFP, the 2A esterase of FMDV, and HOXB4 (FMEV-eGFP-2A-HOXB4+wPRE) [6, 26] (Fig. 1a, b). We prefer the transduction of undifferentiated ES cells because it allows for subsequent isolation of single-cell clones, extensive molecular characterization of the manipulated cells, and selection and expansion of suitable clones prior to their further usage for differentiation followed by in vitro assays or transplantation, in vivo. One disadvantage, however, is that retroviral vectors are often silenced in pluripotent cells necessitating the repeated purification of cultured cells to obtain long-term expressing ES-cell clones. Alternatively, retroviral transduction can also be performed after ES-cell differentiation. In our hands, this also works well for subsequent expansion and analysis of hematopoietic cells, in vitro [17]. However, one should be aware that such polyclonal ES cell-derived hematopoietic cultures which contain many different retroviral integration sites may show biological effects unrelated to HOXB4 expression due to insertional mutagenesis.
5. Because of the short culture times, it is necessary to cryopreserve a large number of cells from a low passage number. For longer growth periods of propagation in the undifferentiated state, it is helpful to maintain CCE cells on growth-arrested murine embryonic fibroblasts (MEFs).
6. We routinely propagate ES cell in 25 m^2 cell culture flasks and passage them every 2 days. Undifferentiated embryonic stem cells will grow as compacted cell aggregates which present themselves under the phase contrast microscope with a smooth, "shiny" appearing border at their edges. Individual cell boundaries usually cannot be recognized (Fig. 2a). The individual colonies should be relatively uniform in size and only few of those should appear as differentiating colonies, which appear flatter with frazzled edges and individual cells being clearly visible (Fig. 2b). Mouse ES-cell colonies grow very fast; thus the cultures should be evaluated every day. If the growth conditions become suboptimal, e.g., acidified "yellow" medium, too large colonies, ES cells will irrevocably differentiate.
7. It is crucial not to overexpose the cells to trypsin but, nonetheless, to get a good single-cell suspension as remaining large aggregates begin to spontaneously differentiate. Colony dissociation can be completed effectively by vigorously pipetting the cell suspension against the bottom of the flask with fresh ES-cell medium.

8. The MOI describes the amount of virus used to infect cells. An MOI of 1 means that each cell is transduced only once, in average. For example, if the titer of your virus is 10^6 GTU/ml supernatant and you wish to transduce 10^5 cells with only 1 GTU/cell (=MOI of 1), you will have to use 100 μl of the virus supernatant. If the MOI should be 10, then you would have to use 1 ml of this supernatant, etc.
9. Because retroviruses are often silenced in pluripotent stem cells, repeated rounds of sorting are likely to be necessary to be able to isolate ES-cell clones expressing the retroviral vector stably and long term.
10. The idea is to somewhat slowly adopt the cells to the basal medium which is used for differentiation 2 days later.
11. For evaluation and electronic documentation of cell numbers, size distribution, and viability, we use a "CASY" Cell-counter model TTC (Roche Applied Science) using a 60 μm capillary. The possibility of electronic documentation of culture quality is very useful during culture of ES cell-derived hematopoietic cells.
12. We use cell strainers from Falcon (BD Falcon #352350), which fit perfectly on top of their 50 ml tubes.
13. It is important to completely remove FCS. Otherwise, many adherent cells will appear during the subsequent culture period, which somehow appear to inhibit growth of hematopoietic cells in suspension.
14. Differentiated cells, presumably primitive macrophages and cell debris, will negatively influence growth of the hematopoietic suspension cultures. We use a "CASY" cell counter (*see* **Note 9**) to evaluate the average cell diameter in our cultures, which is about 8–12 μm and peaks at about 10 μm if everything is fine. If the cell diameter distribution starts to shift (appearing "smeary") towards smaller cell diameters, a Ficoll purification should be performed promptly.
15. Correct compensation is a crucial issue when performing multicolor FACS analysis. We routinely perform the FMO-method [30]. The amounts of fluorochrome-coupled antibodies necessary for staining must be adjusted to the amount of cells used. The amount has to be sufficient to obtain a good signal but low enough to avoid unspecific binding, and, thus, false positive detection. Ideally, each antibody should be titrated on cells known to be positive for the according surface marker. As a rough rule of the thumb, however, one can start with 1 μg of antibody per 10^6 antigen-positive cells.
16. The HOXB4 cultures stain positive for many markers of early definitive hematopoiesis: CD31 (PECAM), a subpopulation

for CD41, CD117 (cKit), and CD45, and many more; for further details *see* Pilat et al. [12]. Most cells in culture are myeloid progenitor cells being double positive for $CD11b^+$/ $GR1^+$ which only give rise to a transient wave of myeloid repopulation after transplantation into appropriate recipient mice. The $CD41^{low}$, $cKit^+$, $CD45^+$ subpopulation consists of mature neutrophilic granulocytes which are constantly produced by $CD41^{high}$ cells in these cultures.

17. May-Grünwald and Giemsa staining solutions were purchased from Sigma-Aldrich (Cat. No. MG500 and GS500, respectively).

Acknowledgments

This work is dedicated to the memory of Wolfram Ostertag (1937–2010) and Hartmut Beug (1945–2011).

As well as Bernd Schiedlmeier, Christopher Baum (Department of Experimental Hematology, Hannover Medical School), and Bernd Giebel and Peter Horn (Institute for Transfusion Medicine, University Hospital Essen) for their steady support and continuing fruitful discussions.

This work was supported by grants of the Deutsche Forschungsgemeinschaft [KL 1311/4-2; KL1311/5-1, cluster of excellence "REBIRTH" (Exc 62)] and by the NIH (grant number RO1-HL073015-06, subaward No. 1000830668).

References

1. Punzel M, Moore KA, Lemischka IR, Verfaillie CM (1999) The type of stromal feeder used in limiting dilution assays influences frequency and maintenance assessment of human long-term culture initiating cells. Leukemia 13:92–97
2. Zhang CC, Kaba M, Ge G, Xie K, Tong W, Hug C, Lodish HF (2006) Angiopoietin-like proteins stimulate ex vivo expansion of hematopoietic stem cells. Nat Med 12:240–245
3. Zhang CC, Kaba M, Iizuka S, Huynh H, Lodish HF (2008) Angiopoietin-like 5 and IGFBP2 stimulate ex vivo expansion of human cord blood hematopoietic stem cells as assayed by NOD/SCID transplantation. Blood 111:3415–3423
4. Antonchuk J, Sauvageau G, Humphries RK (2002) HOXB4-induced expansion of adult hematopoietic stem cells ex vivo. Cell 109:39–45
5. Sauvageau G, Thorsteinsdottir U, Eaves CJ, Lawrence HJ, Largman C, Lansdorp PM, Humphries RK (1995) Overexpression of HOXB4 in hematopoietic cells causes the selective expansion of more primitive populations in vitro and in vivo. Genes Dev 9:1753–1765
6. Schiedlmeier B, Klump H, Will E, Arman-Kalcek G, Li Z, Wang Z, Rimek A, Friel J, Baum C, Ostertag W (2003) High-level ectopic HOXB4 expression confers a profound in vivo competitive growth advantage on human cord blood CD34+ cells, but impairs lymphomyeloid differentiation. Blood 101:1759–1768
7. Challen GA, Boles N, Lin KK, Goodell MA (2009) Mouse hematopoietic stem cell identification and analysis. Cytometry A 75:14–24
8. Giebel B, Punzel M (2008) Lineage development of hematopoietic stem and progenitor cells. Biol Chem 389:813–824

9. Kiel MJ, Yilmaz OH, Iwashita T, Terhorst C, Morrison SJ (2005) SLAM family receptors distinguish hematopoietic stem and progenitor cells and reveal endothelial niches for stem cells. Cell 121:1109–1121
10. McKenzie JL, Takenaka K, Gan OI, Doedens M, Dick JE (2007) Low rhodamine 123 retention identifies long-term human hematopoietic stem cells within the Lin-CD34+CD38- population. Blood 109:543–545
11. Kyba M, Perlingeiro RC, Daley GQ (2002) HoxB4 confers definitive lymphoid-myeloid engraftment potential on embryonic stem cell and yolk sac hematopoietic progenitors. Cell 109:29–37
12. Pilat S, Carotta S, Schiedlmeier B, Kamino K, Mairhofer A, Will E, Modlich U, Steinlein P, Ostertag W, Baum C, Beug H, Klump H (2005) HOXB4 enforces equivalent fates of ES-cell-derived and adult hematopoietic cells. Proc Natl Acad Sci U S A 102:12101–12106
13. Chan KM, Bonde S, Klump H, Zavazava N (2008) Hematopoiesis and immunity of HOXB4-transduced embryonic stem cell-derived hematopoietic progenitor cells. Blood 111:2953–2961
14. Pilat S, Carotta S, Schiedlmeier B, Kamino K, Mairhofer A, Schmidt M, von Kalle C, Steinlein P, Ostertag W, Baum C, Beug H, Klump H (2005) Hematopoietic repopulation, in vivo, with genetically defined clones derived from HOXB4 expressing ES-cells. ASH Annual Meeting Abstracts 106:196
15. Klump H, Schiedlmeier B, Baum C (2005) Control of self-renewal and differentiation of hematopoietic stem cells: HOXB4 on the threshold. Ann N YAcad Sci 1044:1–10
16. Bonde S, Chan KM, Zavazava N (2008) ES-cell derived hematopoietic cells induce transplantation tolerance. PLoS One 3:e3212
17. Schiedlmeier B, Santos AC, Ribeiro A, Moncaut N, Lesinski D, Auer H, Kornacker K, Ostertag W, Baum C, Mallo M, Klump H (2007) HOXB4's road map to stem cell expansion. Proc Natl Acad Sci U S A 104: 16952–16957
18. Takahashi K, Tanabe K, Ohnuki M, Narita M, Ichisaka T, Tomoda K, Yamanaka S (2007) Induction of pluripotent stem cells from adult human fibroblasts by defined factors. Cell 131:861–872
19. Takahashi K, Yamanaka S (2006) Induction of pluripotent stem cells from mouse embryonic and adult fibroblast cultures by defined factors. Cell 126:663–676
20. Wernig M, Meissner A, Foreman R, Brambrink T, Ku M, Hochedlinger K, Bernstein BE, Jaenisch R (2007) In vitro reprogramming of fibroblasts into a pluripotent ES-cell-like state. Nature 448:318–324
21. Park IH, Arora N, Huo H, Maherali N, Ahfeldt T, Shimamura A, Lensch MW, Cowan C, Hochedlinger K, Daley GQ (2008) Disease-specific induced pluripotent stem cells. Cell 134:877–886
22. Kingston RE, Chen CA, Rose JK (2003) Calcium phosphate transfection. Curr Protoc Mol Biol Chapter 9:Unit 9 1
23. Robertson E, Bradley A, Kuehn M, Evans M (1986) Germ-line transmission of genes introduced into cultured pluripotential cells by retroviral vector. Nature 323:445–448
24. Bonde S, Zavazava N (2006) Immunogenicity and engraftment of mouse embryonic stem cells in allogeneic recipients. Stem Cells 24:2192–2201
25. Bonde S, Dowden AM, Chan KM, Tabayoyong WB, Zavazava N (2008) HOXB4 but not BMP4 confers self-renewal properties to ES-derived hematopoietic progenitor cells. Transplantation 86:1803–1809
26. Klump H, Schiedlmeier B, Vogt B, Ryan M, Ostertag W, Baum C (2001) Retroviral vector-mediated expression of HoxB4 in hematopoietic cells using a novel coexpression strategy. Gene Ther 8:811–817
27. Will E, Speidel D, Wang Z, Ghiaur G, Rimek A, Schiedlmeier B, Williams DA, Baum C, Ostertag W, Klump H (2006) HOXB4 inhibits cell growth in a dose-dependent manner and sensitizes cells towards extrinsic cues. Cell Cycle 5:14–22
28. Ferkowicz MJ, Starr M, Xie X, Li W, Johnson SA, Shelley WC, Morrison PR, Yoder MC (2003) CD41 expression defines the onset of primitive and definitive hematopoiesis in the murine embryo. Development 130: 4393–4403
29. Mikkola HK, Fujiwara Y, Schlaeger TM, Traver D, Orkin SH (2003) Expression of CD41 marks the initiation of definitive hematopoiesis in the mouse embryo. Blood 101:508–516
30. Tung JW, Parks DR, Moore WA, Herzenberg LA (2004) New approaches to fluorescence compensation and visualization of FACS data. Clin Immunol 110:277–283
31. Donnelly ML, Luke G, Mehrotra A, Li X, Hughes LE, Gani D, Ryan MD (2001) Analysis of the aphthovirus 2A/2B polyprotein 'cleavage' mechanism indicates not a proteolytic reaction, but a novel translational effect: a putative ribosomal 'skip'. J Gen Virol 82:1013–1025

Chapter 11

Histone Modification Profiling in Normal and Transformed Human Embryonic Stem Cells Using Micro Chromatin Immunoprecipitation, Scalable to Genome-Wide Microarray Analyses

Angelique Schnerch, Shravanti Rampalii, and Mickie Bhatia

Abstract

Comparing normal human embryonic stem cells (hESCs) to those that have acquired cellular properties of neoplasm provides a unique opportunity to study the distinguishing molecular features of human cellular transformation. As global alterations in the epigenetic landscape are a common feature of cancer, we sought to investigate the loci-specific and global differences between normal and transformed hESCs using ChIP-PCR and ChIP-microarray (also known as ChIP-chip). Here, specific emphasis was placed on optimizing ChIP for low cell numbers (termed micro-ChIP; μChIP) towards applications where the target population is rare, such as the case for somatic human tumors containing a low frequency of cancer stem cell populations and for single-colony analysis of embryonic and induced pluripotent stem cells emerging from initial derivation. Using these methods, we suggest that μChIP-PCR and microarray analysis is thus a powerful technology for epigenetic profiling of cell populations relevant to developmental biology, cancer, and regenerative medicine where target populations regulating the biological process can only be isolated in small numbers.

Key words Human embryonic stem cells (hESCs), Induced pluripotent stem cells (iPSC), Transformation, Histone H3 lysine 4 trimethylation, Histone H3 lysine 27 trimethylation, Micro chromatin immunoprecipitation (μChIP), ChIP-PCR, ChIP-chip

1 Introduction

The regulation of the chromatin state is achieved by a number of interrelated mechanisms including covalent histone tail modifications correlated with activation (i.e., trimethylation of histone H3 lysine 4; H3K4me3) and repression (i.e., trimethylation of histone H3 lysine 27; H3K27me3) [1, 2] as well as DNA methylation at CpG islands [3–5]. Epigenetic chromatin modifiers participate in the orchestration of normal stem cell self-renewal and differentiation [6–15] as well as in the process of cellular reprogramming of

Nicholas Zavazava (ed.), *Embryonic Stem Cell Immunobiology: Methods and Protocols*, Methods in Molecular Biology, vol. 1029, DOI 10.1007/978-1-62703-478-4_11, © Springer Science+Business Media New York 2013

somatic cells to a pluripotent stem cell state [16]. Furthermore, aberrant gene expression and an altered global epigenetic state are major features of cancer [17]. Histone modifications (i.e., histone tail methylation) are associated with pathological epigenetic aberrations and poor prognosis, such as in the case of acute myeloid leukemia (AML) [18–20]. Owing to inadequate cell numbers required for global molecular analyses of rare cancer stem cells, our group has defined a model of the cellular and molecular mechanisms of neoplastic transformation in human embryonic stem cells (hESCs) [21, 22] and continues to explore the differences in the epigenetic state between normal and transformed cells in the human. Comparisons of loci-specific and global histone modification patterns in normal and transformed hESCs (thESCs) will be critical to understand fundamental properties of self-renewal and differentiation, particularly how these processes may be dysregulated in cancer stem cells.

Chromatin immunoprecipitation (ChIP) is a widely used method to study how proteins interact with the genome [23], such as how specifically modified histone proteins are distributed at the promoter regions of stem cell regulatory genes and genome-wide. We describe the protocol for ChIP using antibodies against histones H3K4me3 and H3K27me3. This protocol has been optimized for low cell numbers, on the order of 10^4 cells (micro ChIP; μChIP) [24, 25] for PCR analysis or scaled-up for microarray analysis (ChIP-chip) with the incorporation of whole-genome amplification [25, 26]. The μChIP protocol addresses a major limitation of ChIP-chip in which a substantial number of cells (10^7–10^8) are generally required to obtain a robust and reproducible signal. The use of normal and transformed hESCs provides us with the opportunity to optimize the μChIP-chip through comparison with the conventional ChIP-chip protocol and test whether differences identified are truly representative when approaching limiting number of cells isolated for analysis. The optimization of μChIP-chip protocol has widespread applications in the study of exceedingly rare populations of de novo-isolated cancer stem cells and their normal counterparts and in the study of normal hESC differentiation or screening of individual hESC colonies under different pharmacological conditions. An exciting opportunity for ChIP-chip analysis from low cell numbers presents itself in the study of patient-specific induced pluripotent stem cells (iPSC) to model the epigenetic mechanisms of disease pathogenesis. The analysis of the global epigenetic status of individual iPSC colonies is crucial to assess complete reprogramming and ultimately to elucidate the epigenetic changes that occur in subfractions of emerging colonies that are heterogeneous, and could be selected for optimal differentiation potential specific to the desired differentiated cell type to be used for cellular transplantation or drug screening.

2 Materials

2.1 Cell Preparation

1. Mouse embryonic fibroblast-conditioned medium (MEF-CM) supplemented with 8 ng/ml human recombinant basic fibroblast growth factor (hbFGF; Gibco, cat. no. 13256-029). Medium consisted of 80 % knockout Dulbecco modified eagle's medium (KO-DMEM; Gibco, cat. no. 10829-018) supplemented with 20 % knockout serum replacement (KO-SR; Gibco; cat. no. 10828-028), 1 % nonessential amino acids, 1 μM l-glutamine (Gibco, cat. no. 11140-050), *4 ng/ml bFGF*, and 0.1 μM b-mercaptoethanol (Sigma Aldrich, cat. no M7522). MEF-CM was produced over a 7-day period by daily collection of medium used to feed irradiated (40 Gy) MEFs (G1:CF-1^{R}BR; Charles River Canada, St-Constant, Quebec, Canada) [27].
2. Human recombinant basic fibroblast growth factor (hbFGF) is dissolved at 1 μg/ml in D-PBS (Gibco, cat. no. 14190-144) containing 0.1 % BSA, filtered (pretreat filter with 10 % BSA solution—1 ml BSA+2 ml PBS), stored in aliquots at −80 °C, and added to medium as required (*see* **Note 1**).
3. Matrigel (BD Biosciences, cat. no. 353234). Matrigel aliquots are prepared *on ice* as follows: Thaw Matrigel overnight at 4 °C to ensure that a gel does not form. Add 10 ml of ice-cold KO-DMEM directly to bottle containing 10 ml of Matrigel and mix well with a 10 ml serological pipette. Aliquot 1 or 2 ml *on ice* into *chilled* 15 ml falcon tubes (BD Biosciences, cat. no. 35209). Store at −30 °C.
4. TrypLE™ Express Stable Trypsin Replacement Enzyme without Phenol Red (Invitrogen, cat. no. 12604-013). Stored at 4 °C, protected from light.
5. Dulbecco's phosphate buffered saline (D-PBS), 1× (Gibco, 14190-144).
6. Trypan blue stain 0.4 % (Invitrogen, cat. no. T10282).

2.2 Chromatin Immunoprecipitation

1. ChIP-grade anti-histone antibodies: Anti-H3K4me3 (Abcam, cat. no. ab1012) and anti-H3K27me3 (Abcam, cat. no. ab6002).
2. Normal mouse serum (Santa Cruz, cat. no. sc-45051).
3. Dynabeads Protein G for immunoprecipitation (Invitrogen, cat. no. 100-03D) (vortex the beads immediately before use to ensure that the beads are well suspended) (*see* **Note 2**).
4. Siliconized Low-Retention Microcentrifuge Tubes (Fisher, cat. no. 02-681-331).
5. 36.5–38 % Formaldehyde (Sigma-Aldrich, cat. no. F8775) (*see* **Note 3**).

6. Glycine (Sigma-Aldrich, cat. no. D5760): Make up a 1.25 M stock solution in water (powder molecular weight 75.07 g/mol; use 9.385 g and add water up to 100 ml). Prior to use, autoclave stock solution and store at RT.
7. Complete EDTA-free, Protease Inhibitor Cocktail Tablets (Roche, cat. no. 11 873 580 001).
8. DynaMag™-2 magnet (Invitrogen, cat. no. 123-21D).
9. Lysis buffer—50 mM Tris–HCL, pH 8.0, 10 mM EDTA, 1 % SDS, protease inhibitor cocktail diluted 1:100 (one tablet dissolved in 1 ml H_2O stock solution), and 1 mM PMSF. Protease inhibitor cocktail and PMSF must be added freshly before use: this constitutes the *complete lysis buffer*.
10. RIPA buffer—10 mM Tris–HCL, pH 7.5, 140 mM NaCL, 1 mM EDTA, 0.5 mM EGTA, 1 % Triton-X, SDS (0.1 %, w/v), Na-deoxycholate (0.1 %, w/v). Store at 4 °C protected from the light.
11. RIPA ChIP buffer—10 mM Tris–HCl, pH 7.5, 140 mM NaCL, 1 mM EDTA, 0.5 mM EGTA, 1 % Triton-X, SDS (0.1 %, w/v), Na-deoxycholate (0.1 %, w/v), protease inhibitor cocktail diluted 1:100 from stock solution, and 1 mM PMSF. Protease inhibitor cocktail and PMSF are added fresh before use. Store at 4 °C protected from the light.

2.3 Whole-Genome Amplification

1. GenomePlex® Single Cell Whole Genome Amplification Kit (Sigma, cat. no. WGA4). Store at −20 °C.
2. GenomePlex® WGA Reamplification Kit (Sigma, cat. no. WGA3). Store at −20 °C.
3. Sigma-Aldrich GenElute PCR Clean-up Kit (Sigma-Aldrich, cat. no. NA1020). Columns should be stored at 4 °C; remainder of reagents can be kept at room temperature.

2.4 Quantitative Polymerize Chain Reaction

1. Platinum SYBR Green qPCR SuperMix-UDG, store at −20 °C (Invitrogen; cat. no. 11733-046).

3 Methods

3.1 Preparation of Antibody–Magnetic Bead Complexes for mChIP Assay

1. Prepare slurry of Dynabeads Protein G. For each IP, 11 μl of beads are required. For six ChIPs (two IPs and one negative control per sample, plus one extra aliquot in case of loss/pipetting error) take 77 μl of suspended Dynabeads Protein G into a 1.5-ml tube and place in the magnetic holder, allow beads to be captured (~1 min), remove tubes from the magnet, and wash with 154 μl of ice-cold RIPA buffer (22 μl per initial 11 μl of beads used).

2. Vortex, then capture the beads using the magnet, remove the buffer, and add 77 μl of RIPA buffer.
3. Vortex the beads and place on ice.
4. Aliquot 90 μl RIPA buffer into 1.5 ml tubes for each ChIP reaction (six tubes).
5. Add 10 μl of prewashed Dynabeads Protein G per tube.
6. Add 3–5 μg specific antibody (anti-H4K4me3 or anti-H3K27me3) to each tube and 1.5 μl normal mouse serum per tube for the hESC and thESC negative controls.
7. Incubate the tubes on a rotator set to 40 rpm for a minimum of 2 h at RT or overnight at 4 °C.

3.2 Preparation of Samples for ChIP Assay

1. Normal and transformed hESCs are harvested at day 7 when cultures are confluent using 1 ml of pre-warmed tripLE[1] (6-well tissue culture plates) for 5 min (*see* **Note 4**). Cultures are rinsed once in the plate with 1 ml dPBS, add 1 ml dPBS, scrape, triturate gently ~20 times (until single-cell suspension is achieved), and transfer to a 1.5 ml tube. Take an aliquot of normal and transformed hESC to perform cell counts using trypan blue exclusion method (*see* **Note 5**). Remove the appropriate volume of cell suspension required to obtain 10,000 cells per IP for each sample, transfer to a new 1.5 ml siliconized Eppendorf tube, and bring the total volume of cell suspension to 1 ml with 1× dPBS.

3.3 Micro Chromatin Immunoprecipitation Assay

1. Cross-link cells by adding 27 μl of formaldehyde to a final concentration of 1 %, mix by gentle vortexing, and rotate at room temperature for 8–10 min.
2. Stop cross-linking by adding 114 μl of 1.25 M glycine at a final concentration of 125 mM, gently vortex, and rotate at room temperature for 5 min.
3. Spin down the cells at 1,500 rpm for 10 min at 4 °C (*see* **Note 6**).
4. Carefully aspirate the supernatant leaving behind approximately 30 μl of the solution containing the cell pellet to ensure maximum recovery of the cells, as the pellet will not be visible.
5. Rinse the cells with 500 μl PBS and centrifuge cells at 1,500 rpm for 10 min at 4 °C.
6. Aspirate the PBS leaving behind 30 μl of the solution plus cell pellet (*see* **Note 7**).
7. Add 120 μl of room-temperature *complete lysis buffer* to the cells, resuspend by gently pipetting up and down, and incubate for 5 min on ice.

---

[1]Cells are cultured in 6-well tissue culture plates and passaged at a ratio of 1:2 upon reaching confluency every 7 days using pre-warmed Collagenase IV. A working solution of 0.5 ml Collagenase IV.

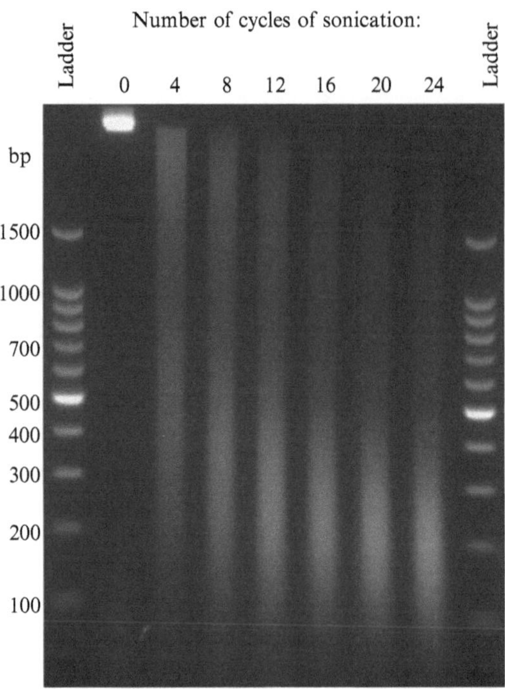

Fig. 1 Optimization of sonication regimes to obtain chromatin fragments from approximately 400–500 bp, which is optimal for ChIP assays as they span two to three nucleosomes. The number of sonication cycles will vary greatly depending on cell type, degree of cross-linking, reaction volume, and the specific instrument used. Undersonication will result in a loss of resolution while oversonication will increase noise. (**a**) Example of agarose gel electrophoresis assessment of cross-linked chromatin shearing by running a time course of increasing sonication cycles (adapted from Lee et al. [28]). Cross-links were reversed; the DNA was purified, and resolved in 2 % agarose. Lanes with molecular weight are labeled ladder and lanes with sheared chromatin are labeled with the number of sonication cycles

8. Sonicate the cells using the Bioruptor; we specifically used 17.5 cycles of [30-s "ON", 60-s "OFF"] each to obtain fragment sizes of approximately 400 bp (*see* **Notes 8** and **9**) (*see* Fig. 1).
9. Add 870 μl of RIPA ChIP buffer to the sheared chromatin (*see* **Note 10**).
10. Centrifuge at 12,000×*g* for 10 min at 4 °C, and take the supernatant containing the chromatin, leaving behind a residual ~30–50 μl in the tube. This residual amount contains cellular debris, which may adhere to the magnetic beads and cause unspecific background.
11. Remove 100 μl for input, and store at 4 °C.

12. Take the antibody–bead-containing tubes, place in the magnetic rack for 1 min, and remove the supernatant (*see* **Note 11**). Wash once with RIPA buffer and equally aliquot the beads into fresh tubes.
13. Evenly distribute the sheared chromatin from normal and transformed hESCs to the antibody–bead-containing tubes (~300 μl per ChIP; recall three ChIPs per sample).
14. Remove the tubes from the magnetic rack and place the tubes on a rotator set at 40 rpm for a minimum 2 h up to overnight at 4 °C.
15. Place the tubes in the magnetic rack; remove liquid caught in the lid as per **Note 11**. The chromatin–antibody–bead (immune) complexes are captured on the wall of the tube.
16. Discard the supernatant.
17. Wash three times using 100 μl of ice-cold RIPA buffer. Complete each wash as follows: Add buffer, hand vortex or invert to mix, incubate for 4 min on a rotator (40 rpm) at 4 °C, place in the magnetic rack, remove liquid caught in lid according to **Note 11**, wait for 1 min, discard the supernatant, and keep the beads.
18. Wash one time with 100 μl TE buffer, incubate for 4 min on a rotator (40 rpm) at 4 °C, remove liquid caught in lid by spinning down in a microcentrifuge for 1 s, and transfer contents to a *new*, labeled 1.5 ml tube.
19. Place tubes in magnetic rack, discard the TE, remove from magnetic rack, and keep the beads on ice.
20. Add 200 μl elution buffer to the immune complexes and input samples from step 2.3.11.
21. Vortex and incubate for 5 min at 65 °C.
22. Spin down tubes for 1 s in a microcentrifuge, place tubes in magnetic rack for 1 min, and transfer supernatant (Elution 1, E1) to a new, labeled tube.
23. Add 200 μl elution buffer to all tubes.
24. Vortex and incubate for 5 min at 65 °C.
25. Spin down tubes for 1 s in a microcentrifuge, place tubes in magnetic rack for 1 min, collect the supernatant (Elution 2, E2), and combine it with E1.
26. Add 16 μl of 5 M NaCl (these calculations are for 400 μl eluate; add 8 ml NaCL and 2 ml Rnase A to input samples) and 4 μl of Rnase A to each tube.
27. Incubate for 5 h at 65 °C.
28. Spin down tubes for 1 s and add 4 μl EDTA and 1 μl proteinase K.

29. Incubate for 2 h at 42 °C.
30. Purify the DNA in a QIAquick MINelute column (follow the instructions for MinElute Reaction Cleanup Kit) (*see* **Note 12**). Elute twice with 20 μl of kit provided Buffer EB (elution buffer). Measure concentration by NanoDrop.

3.4 Amplicon Preparation Using GenomePlex® Single Cell Whole Genome Amplification Kit

1. Use 11 μl of purified DNA from **step 30** of Subheading 3.3 for first-round linear amplification.
2. Add 2 μl of *1× Library Preparation Buffer* to each sample.
3. Add 1 μl of *Library Stabilization Solution.*
4. Vortex, centrifuge for 1 s, and place in thermal cycler for 2 min at 95 °C, followed by 1 min at 4 °C.
5. Centrifuge for 1 s, and place samples on ice.
6. Add 1 μl *Library Preparation Enzyme*, vortex, and centrifuge for 1 s.
7. Place samples in a thermal cycler and incubate as follows: 16 °C for 20 min; 24 °C for 20 min; 37 °C for 20 min; 75 °C for 5 min; and 4 °C hold.
8. Centrifuge for 1 s. Samples may be amplified immediately or stored at −20 °C for up to 3 days.
9. Amplification master mix may be prepared by adding the following reagents to the 15 μl reaction from step 2.4.8: 7.5 μl of *10× Amplification Master Mix*, 46.675 μl of nuclease-free water, 0.11 mM dUTP (0.825 μl of 10× stock dUTP—10 mM in 22 μl nuclease-free water) (*see* **Note 13**), and 5 μl of *WGA DNA polymerase.*
10. Vortex, centrifuge for 1 s, and begin thermal cycling:

    Initial denaturation: 95 °C for 3 min.

    Perform 14 cycles as follows:

        Denature: 94 °C for 30 s.

        Anneal/extend: 65 °C for 5 min.

    Maintain reactions at 4 °C or store at −20 °C.
11. Clean up reactions using Sigma PCR clean-up kit as per the manufacturer's instructions (*see* **Note 14**). Elute in 50 μl of kit provided *Elution Solution.* Measure concentration by NanoDrop.

3.5 Confirmation of WGA4 Amplicons by ChIP-PCR

1. Dilute amplicons to 10 ng/μl and test 5 μl by quantitative PCR with control primer sets targeting OCT4 and Brachyury promoter regions. An example of the results is shown in Fig. 2.
2. Use 5 μl of template for qPCR using the Platinum® SYBR® Green qPCR SuperMix-UDG and following the manufacturer's instructions (*see* **Note 15**).

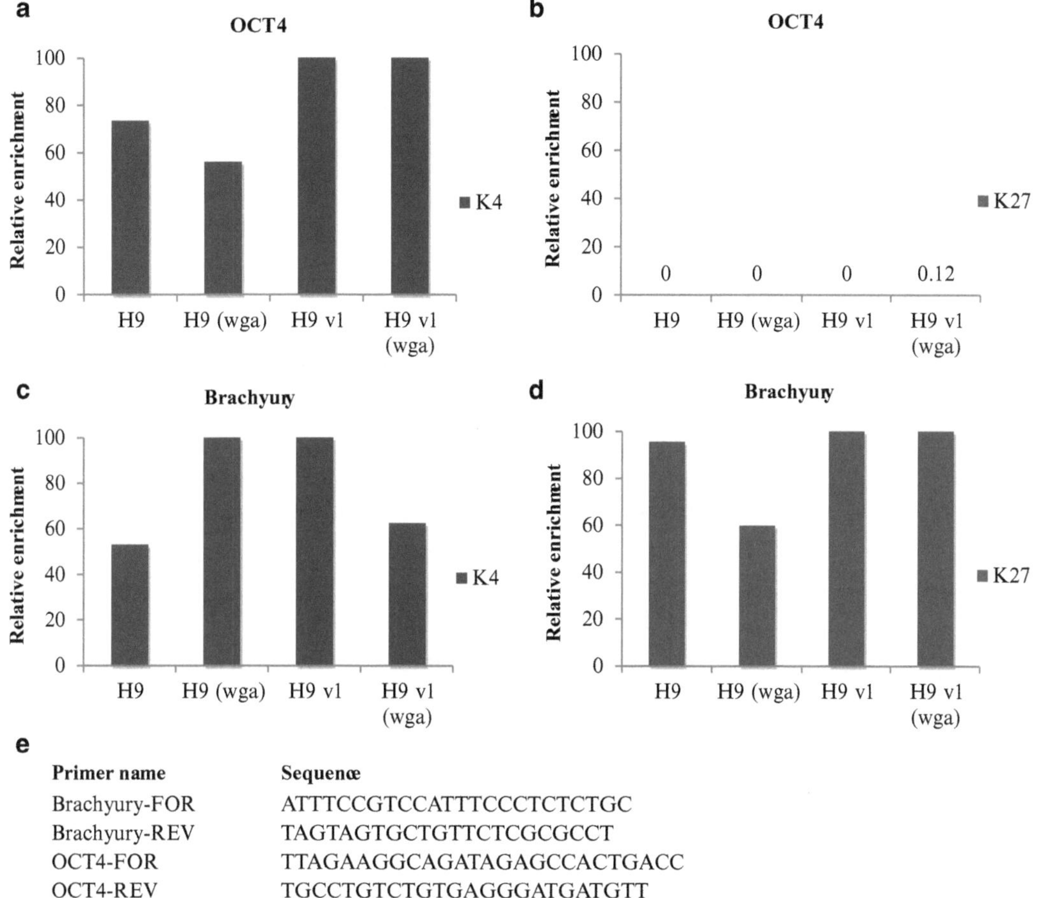

Primer name	Sequence
Brachyury-FOR	ATTTCCGTCCATTTCCCTCTCTGC
Brachyury-REV	TAGTAGTGCTGTTCTCGCGCCT
OCT4-FOR	TTAGAAGGCAGATAGAGCCACTGACC
OCT4-REV	TGCCTGTCTGTGAGGGATGATGTT

Fig. 2 ChIP-qPCR results from whole-genome amplicons compared with un-amplified samples from normal and transformed hESCs. (**a**, **b**) Enrichment of H3K4me3 and H3K27me3 modifications was assayed at the promoter regions of the pluripotency gene OCT4 and (**c**, **d**) mesoderm-specific gene Brachyury. (**e**) Table of primer sequences for OCT4 and Brachyury promoters

3.6 Reamplification of WGA4 Amplicons for Microarray Scale-Up Using the Genomeplex® WGA Reamplification Kit

1. Add 10 ng of 1 ng/µl WGA4 amplicons and amplify using WGA3 kit. Follow the manufacturer's instructions (Reamplification Procedure A). Amplification mix: To 10 µl of WGA4 amplicons add:

   48.675 µl of nuclease-free water.

   7.5 µl of 10× Amplification Master Mix.

   3.0 µl of the 10 nM dNTP mix.

   0.825 µl of dUTP.

   5 µl of WGA DNA Polymerase.

2. Clean up reactions using Sigma PCR clean-up kit as per the manufacturer's instructions. Elute in 50 µl of kit provided *Elution Solution*. Measure concentration by NanoDrop.

3.7 Confirmation of WGA3 Amplicons

1. Dilute amplicons to 10 ng/μl and test 5 μl by quantitative PCR with control primer sets targeting OCT4 and Brachyury promoter regions.
2. Use 5 μl of template for qPCR using the Platinum® SYBR® Green qPCR SuperMix-UDG (*see* **Note 15**).

4 Notes

1. Add 8 ng/ml of hbFGF to MEF-CM prior to use.
2. Use Dynabeads Protein G with mouse IgGs and Dynabeads Protein A (Invitrogen, cat. no. 100-01D) with rabbit IgGs.
3. Formaldehyde is toxic by inhalation, and physical contact, or if ingested and should be disposed of according to standard operating protocols for hazardous waste.
4. The ChIP protocol should be used for 1×10^5–1×10^6 live cells per immunoprecipitation while the mChIP protocol should be used for 1×10^4 live cells.
5. For the trypan blue exclusion method dilute the cells to 1–2×10^5 cells/ml, and for a single well of a 6-well tissue culture plate the dilution is approximately 1:10. To 10 μl of cell suspension, add 10 μl of trypan blue stain, mix thoroughly, and incubate at room temperature for 5 min prior to measurement of live/dead cells using a hemocytometer or an automated cell counter (Countess Cell Counter; Invitrogen, cat. no. C10227).
6. At this point all steps are carried out on ice, try to chill magnetic stand and all buffers/solutions (unless otherwise stated), and precool microcentrifuge tubes used in subsequent steps.
7. Cells can be stored at −80 °C at this point for later use.
8. Bioruptor is operated in a 4 °C refrigerator. The reservoir is filled to the water-level mark with slurry of 4 °C water and ice chips (70:30 ratio). The Bioruptor is set to "High Power." Replace the gradually heated water with new water/ice chip slurry every five cycles (keep samples on ice during this procedure).
9. A sonication curve should be generated using hESCs to determine the optimal number of cycles needed to generate chromatin fragment sizes of ~400–600 bp.
10. Additional RIPA buffer is added after sonication to reduce the concentration of SDS to approximately 0.1 %.
11. To remove liquid caught in the lid of the tube during rotation, place tube in the magnetic rack for a moment, invert rack (containing tubes) for an instant, and quickly bring rack to original position.

12. Protocol for cleanup reaction is as follows:
    (a) Add 3 volumes of buffer ERC to the enzymatic reaction and mix. The maximum volume of enzymatic reaction that can be processed per MinElute column is 100 μl; if the reaction volume exceeds 100 μl split the sample and use the appropriate number of columns.
    (b) Check that the color of the mixture is yellow (similar to buffer ERC without the enzymatic reaction). If the color of the mixture is orange or violet, add 10 μl of 3 M sodium acetate, pH 5.0, and mix. The color of the mixture will turn to yellow.
    (c) Place a MinElute column in a 2 ml collection tube, apply the sample to the MinElute column, and centrifuge for 1 min. Discard the eluate and replace the column into the collection tube.
    (d) Wash with 750 μl buffer PE added to the MinElute column and centrifuge for 1 min. Note: Be sure to add ethanol to the concentrated wash buffer.
    (e) Discard the flow-through and place the MinElute column back in the same tube. Centrifuge the column for an additional 1 min at maximum speed.
    (f) Place the MinElute column in a clean 1.5 ml microcentrifuge tube.
    (g) To elute DNA, add 10 μl buffer EB (10 mM Tris-HCl, pH 8.5) to the center of the membrane, let the column stand for 1 min, and then centrifuge for 1 min. Pool all eluates.
13. Addition of dUTP is required for ChIP-on-chip applications using Affymetrix promoter or whole-genome tiling arrays. Necessary to the Affymetrix protocol is enzymatic fragmentation and labeling of DNA fragments dependent on random incorporation of dUTP during whole-genome amplification prior to array hybridization.
14. All centrifugations are done at maximum speed (12,000–16,000 × *g*). The protocol is as follows:
    (a) Insert a GenElute Miniprep Binding Column (with a blue o-ring) into a provided collection tube. Add 0.5 ml of the column preparation solution to each miniprep column and centrifuge for 30 s. Discard eluate.
    (b) Add 5 volumes of binding solution to 1 volume of the PCR reaction and mix. For example, add 375 μl of binding solution to 75 μl of the WGA reaction. Transfer the solution into the binding column. Centrifuge the column at maximum speed for 1 min. Discard the eluate, but retain the collection tube.

(c) Replace the binding column into the collection tube. Apply 0.5 ml of diluted wash solution to the column and centrifuge at maximum speed for 1 min. Discard the eluate, but retain the collection tube. Note: Be sure to add ethanol to the wash solution concentrate prior to first-time use.

(d) Replace the column into the collection tube. Centrifuge the column at maximum speed for 2 min, to remove excess ethanol. Discard any residual eluate as well as the collection tube.

(e) Transfer the column to a fresh 2 ml collection tube. Apply 50 μl of elution solution to the center of each column. Incubate at room temperature for 1 min.

(f) To elute the DNA, centrifuge the column at maximum speed for 1 min. Store at −20 °C.

15. Standard protocol is provided for 50 μl reaction size, scale-down qPCR reaction components to 25 μl total reaction volume. Briefly, prepare a master-mix of all components and save for the template to reduce pipetting error. The components per reaction are as follows: 12.5 μl of Platinum Green qPCR SuperMix-UDG, 1 μl of gene-specific forward and reverse primers (10 μM working dilution), and 6.5 μl of water. Keep the master-mix chilled and light protected. Add 20 μl of the master-mix per each well in a 96-well plate, to which 5 μl of template is added. Samples are then incubated as follows:

50 °C for 2-min hold.

95 °C for 2-min hold.

40 Cycles of:

95 °C, 15 s.

60 °C, 30 s.

Followed by melting curve analysis: Refer to instrument documentation.

References

1. Jenuwein T, Allis CD (2001) Translating the histone code. Science 293:1074–1080
2. Strahl BD, Allis CD (2000) The language of covalent histone modifications. Nature 403: 41–45
3. Bird AP, Wolffe AP (1999) Methylation-induced repression—belts, braces, and chromatin. Cell 99:451–454
4. Jones PA, Takai D (2001) The role of DNA methylation in mammalian epigenetics. Science 293:1068–1070
5. Fuks F (2005) DNA methylation and histone modifications: teaming up to silence genes. Curr Opin Genet Dev 15:490–495
6. Boyer LA, Plath K, Zeitlinger J, Brambrink T, Medeiros LA, Lee TI, Levine SS, Wernig M, Tajonar A, Ray MK, Bell GW, Otte AP, Vidal M, Gifford DK, Young RA, Jaenisch R (2006) Polycomb complexes repress developmental regulators in murine embryonic stem cells. Nature 441:349–353
7. Bracken AP, Dietrich N, Pasini D, Hansen KH, Helin K (2006) Genome-wide mapping of polycomb target genes unravels their roles in cell fate transitions. Genes Dev 20:1123–1136
8. Endoh M, Endo TA, Endoh T, Fujimura Y, Ohara O, Toyoda T, Otte AP, Okano M,

Brockdorff N, Vidal M, Koseki H (2008) Polycomb group proteins Ring1A/B are functionally linked to the core transcriptional regulatory circuitry to maintain ES cell identity. Development 135:1513–1524

9. Ernst P, Fisher JK, Avery W, Wade S, Foy D, Korsmeyer SJ (2004) Definitive hematopoiesis requires the mixed-lineage leukemia gene. Dev Cell 6:437–443
10. Ernst P, Mabon M, Davidson AJ, Zon LI, Korsmeyer SJ (2004) An Mll-dependent Hox program drives hematopoietic progenitor expansion. Curr Biol 14:2063–2069
11. Ernst P, Wang J, Korsmeyer SJ (2002) The role of MLL in hematopoiesis and leukemia. Curr Opin Hematol 9:282–287
12. Kajiume T, Ninomiya Y, Ishihara H, Kanno R, Kanno M (2004) Polycomb group gene mel-18 modulates the self-renewal activity and cell cycle status of hematopoietic stem cells. Exp Hematol 32:571–578
13. Lee TI, Jenner RG, Boyer LA, Guenther MG, Levine SS, Kumar RM, Chevalier B, Johnstone SE, Cole MF, Isono K, Koseki H, Fuchikami T, Abe K, Murray HL, Zucker JP, Yuan B, Bell GW, Herbolsheimer E, Hannett NM, Sun K, Odom DT, Otte AP, Volkert TL, Bartel DP, Melton DA, Gifford DK, Jaenisch R, Young RA (2006) Control of developmental regulators by Polycomb in human embryonic stem cells. Cell 125:301–313
14. Mikkelsen TS, Ku M, Jaffe DB, Issac B, Lieberman E, Giannoukos G, Alvarez P, Brockman W, Kim TK, Koche RP, Lee W, Mendenhall E, O'Donovan A, Presser A, Russ C, Xie X, Meissner A, Wernig M, Jaenisch R, Nusbaum C, Lander ES, Bernstein BE (2007) Genome-wide maps of chromatin state in pluripotent and lineage-committed cells. Nature 448:553–560
15. Park IK, Qian D, Kiel M, Becker MW, Pihalja M, Weissman IL, Morrison SJ, Clarke MF (2003) Bmi-1 is required for maintenance of adult self-renewing haematopoietic stem cells. Nature 423:302–305
16. Mikkelsen TS, Hanna J, Zhang X, Ku M, Wernig M, Schorderet P, Bernstein BE, Jaenisch R, Lander ES, Meissner A (2008) Dissecting direct reprogramming through integrative genomic analysis. Nature 454: 49–55
17. Jones PA, Baylin SB (2007) The epigenomics of cancer. Cell 128:683–692
18. Fraga MF, Ballestar E, Villar-Garea A, Boix-Chornet M, Espada J, Schotta G, Bonaldi T, Haydon C, Ropero S, Petrie K, Iyer NG, Perez-Rosado A, Calvo E, Lopez JA, Cano A, Calasanz MJ, Colomer D, Piris MA, Ahn N, Imhof A, Caldas C, Jenuwein T, Esteller M (2005) Loss of acetylation at Lys16 and trimethylation at Lys20 of histone H4 is a common hallmark of human cancer. Nat Genet 37:391–400
19. Seligson DB, Horvath S, Shi T, Yu H, Tze S, Grunstein M, Kurdistani SK (2005) Global histone modification patterns predict risk of prostate cancer recurrence. Nature 435:1262–1266
20. Eguchi M, Eguchi-Ishimae M, Greaves M (2005) Molecular pathogenesis of MLL-associated leukemias. Int J Hematol 82:9–20
21. Ji J, Werbowetski-Ogilvie TE, Zhong B, Hong SH, Bhatia M (2009) Pluripotent transcription factors possess distinct roles in normal versus transformed human stem cells. PLoS One 4:e8065
22. Werbowetski-Ogilvie TE, Bosse M, Stewart M, Schnerch A, Ramos-Mejia V, Rouleau A, Wynder T, Smith MJ, Dingwall S, Carter T, Williams C, Harris C, Dolling J, Wynder C, Boreham D, Bhatia M (2009) Characterization of human embryonic stem cells with features of neoplastic progression. Nat Biotechnol 27:91–97
23. Orlando V (2000) Mapping chromosomal proteins in vivo by formaldehyde-crosslinked-chromatin immunoprecipitation. Trends Biochem Sci 25:99–104
24. Dahl JA, Collas P (2008) A rapid micro chromatin immunoprecipitation assay (microChIP). Nat Protoc 3:1032–1045
25. Acevedo LG, Iniguez AL, Holster HL, Zhang X, Green R, Farnham PJ (2007) Genome-scale ChIP-chip analysis using 10,000 human cells. Biotechniques 43:791–797
26. O'Geen H, Nicolet CM, Blahnik K, Green R, Farnham PJ (2006) Comparison of sample preparation methods for ChIP-chip assays. Biotechniques 41:577–580
27. Chadwick K, Wang L, Li L, Menendez P, Murdoch B, Rouleau A, Bhatia M (2003) Cytokines and BMP-4 promote hematopoietic differentiation of human embryonic stem cells. Blood 102:906–915
28. Lee TI, Johnstone SE, Young RA (2006) Chromatin immunoprecipitation and microarray-based analysis of protein location. Nat Protoc 1:729–748

Chapter 12

Combined Total Proteomic and Phosphoproteomic Analysis of Human Pluripotent Stem Cells

Junjie Hou, Brian T.D. Tobe, Frederick Lo, Justin D. Blethrow, Andrew M. Crain, Dieter A. Wolf, Evan Y. Snyder, Ilyas Singec, and Laurence M. Brill

Abstract

Despite advances in understanding pluripotency through traditional cell biology and gene expression profiling, the signaling networks responsible for maintenance of pluripotency and lineage-specific differentiation are poorly defined. To aid in an improved understanding of these networks at the systems level, we present procedures for the combined analysis of the total proteome and total phosphoproteome (termed (phospho) proteome) from human embryonic stem cells (hESCs), human induced pluripotent stem cells (hiPSCs), and their differentiated derivatives. Because there has been considerable heterogeneity in the literature on the culture of pluripotent cells, we first briefly describe our feeder-free cell culture protocol. The focus, however, is on procedures necessary to generate large-scale (phospho)proteomic data from the cells. Human cells are described here, but the (phospho)proteomic procedures are broadly applicable. Detailed procedures are given for lysis of the cells, protein sample preparation and digestion, multidimensional liquid chromatography, analysis by tandem mass spectrometry, and database searches for peptide/protein identification (ID). We summarize additional data analysis procedures, the subject of ongoing efforts.

Key words Embryonic stem cells (ESCs), Induced pluripotent stem cells (iPSCs), Mass spectrometry (MS), Multidimensional liquid chromatography (MDLC), Proteomics, Phosphoproteomics, Self-renewal, Pluripotency, Protein phosphorylation, Posttranslational modification (PTM)

1 Introduction

Human pluripotent stem cells, including embryonic stem cells (hESCs) and induced pluripotent stem cells (hiPSCs), have the capacity for potentially unlimited self-renewal and differentiation. Pluripotency is maintained during symmetric cell division through

Authors Junjie Hou and Brian T.D. Tope contributed equally to this chapter.

Nicholas Zavazava (ed.), *Embryonic Stem Cell Immunobiology: Methods and Protocols*, Methods in Molecular Biology, vol. 1029, DOI 10.1007/978-1-62703-478-4_12, © Springer Science+Business Media New York 2013

production of identical, primordial cells, which have the potential to differentiate into more than 200 different cell types of the human body [1, 2]. Pluripotent stem cells are therefore a powerful experimental model system in developmental biology and could have therapeutic and diagnostic value for many illnesses of diverse organ systems and tissues [1, 3].

To begin to understand the systems-level basis of pluripotency networks, mRNA expression profiles in mouse embryonic stem cells (mESCs) and hESCs have been examined extensively [4, 5]. However, changes in mRNA expression often do not reflect changes in protein abundance [6], which is also affected by translational regulation, posttranslational modifications (PTMs), and protein degradation. Reversible posttranslational phosphorylation of serine, threonine, and tyrosine residues is the most important known molecular event controlling cell signaling, affecting protein activity, stability, function, complex formation, and localization [7–9].

For the analysis of protein abundance and phosphorylation, multidimensional liquid chromatography-tandem mass spectrometry (MDLC-MS/MS) is the current technology of choice, because it can provide quantitative protein identifications (IDs) and site-specific PTM identifications with high sensitivity on a large scale [10]. In recent years, technologies and methodologies for MDLC-MS/MS-based proteomic analyses have advanced rapidly, including HPLC-based methods for separation of complex mixtures of peptides, phosphopeptide enrichment, and MS/MS analyses of (phospho)peptides. Recent advances in detection sensitivity, scan speed, mass accuracy, and resolution of mass spectrometers enable the identification of increased number of proteins with high confidence, especially low-abundance (phospho)proteins, with relative and absolute quantification. Here, we outline our pluripotent cell culture methods but focus on the description of a robust, combined proteomic and (phospho)proteomic workflow, built upon but far surpassing one which was published recently [11]. This broadly applicable technology platform has been successfully applied to analyze the total proteome and total (phospho)proteome of cancer cell lines, hESCs, their differentiated derivatives, and entire organs from mice.

2 Materials

We list specific instruments and products used in these protocols, but exclusion of other high-quality substitute products and instruments is NOT intended or implied.

2.1 Cell Culture

1. Light microscope equipped with phase-contrast optics (e.g., Olympus CKX41).
2. Dissecting microscope (Olympus SZ).

3. Sterile laminar-flow tissue culture (TC) hood (Esco).
4. Mouse embryonic fibroblast (MEF) line (CD1 used in this protocol).
5. Human pluripotent stem cells (hESCs or hiPSCs).
6. Mouse embryonic fibroblast-conditioned media (MEF-CM) (*see* **Note 1**).
7. FBS Hyclone, heat inactivated (Invitrogen/Gibco, cat. # SH30071.03HI).
8. Penicillin/Streptomycin (Pen/Strep; Omega Scientific, cat. # PS20).
9. Normocin (Invitrogen, Ant-NR-2).
10. Y-27632 dihydrochloride (Rho-associated kinase (ROCK) inhibitor), 10 mM stock (Tocris, cat. # 1254).
11. FGF2, 20 mg/ml stock (R&D Systems, cat. # 233-FB).
12. Matrigel (Geltrex) (BD Biosciences, cat. # 356237).
13. Cell culture flasks T75 vent cap (Corning, cat. # 430641).
14. Tissue grade 6-well plates (Corning, cat. # 3516).
15. Pipettes, 5, 10 ml (Sarstedt, cat. # 861253001, 861254001).
16. 2 ml aspirating pipette (BD Vacutainer Labware Medical, cat. # 357558).
17. Sterile conical tube, 50 ml (Corning, cat. # 430828).
18. Micro-centrifuge tubes, 1.5 ml (Sarstedt, part # 72.692.005).
19. Cell scrapers (Corning costar, 3010).

2.2 Lysis of hESCs, hiPSCs, or Derivatives for Protein Preparation

Lysis buffer is prepared in high-purity water (HPLC Grade; Chromasolve Plus, Sigma, cat. # 34877-4L) with the following composition (*see* **Note 2**):

Salts, detergents, and chelator stocks	Final concentration
1. 1.0 M Tris pH 7.5	50 mM
2. 5.0 M NaCl	100 mM
3. NP40 (Igepal, CA630 Sigma part # I3021; *see* **Note 3**)	1.0 %
4. 100 % Glycerol	10.0 %
5. 0.5 M EDTA	1.0 mM
Protease and protein phosphatase inhibitors (add immediately before use; *see* **Note 4**)	
6. Sodium orthovanadate crystals	4.0 mM
7. β-Glycerophosphate disodium salt hydrate (Sigma, cat. # G9891)	20 mM

(continued)

Salts, detergents, and chelator stocks	Final concentration
8. Phosphatase inhibitor cocktail 1 stock (Sigma, cat. # P2850; *see* **Note 4**)	1:100
9. Phosphatase inhibitor cocktail 2 stock (Sigma, cat. # P5726)	1:100
10. Phosphatase inhibitor cocktail 3 stock (Sigma, cat. # P0044)	1:100
11. Protease inhibitor cocktail (Sigma, cat. # P8340)	1:100
12. Calyculin A (AG Sciences, prod. # C-1031)	100 nM
13. Pefabloc SC powder (Sigma, cat. # 76307)	2.0 mM

2.3 Ammonium Sulfate ($(NH_4)_2SO_4$) Precipitation, Protein Storage

1. $(NH_4)_2SO_4$ (Sigma, cat. # 09982).

2.4 Protein Resuspension, Gel Filtration, and Quantification

1. PD-10 gel filtration columns (GE Healthcare; cat. # 17-0851-01).
2. 8 M urea/100 mM NH_4HCO_3 (urea, NH_4HCO_3 from Sigma or other high-grade suppliers).
3. 80 % H_2O/20 % glycerol (from Sigma or other high-grade suppliers).
4. Phosphatase inhibitor cocktail 1 stock 1:100 (*see* **Note 4**).
5. Phosphatase inhibitor cocktail 2 stock 1:100.
6. Phosphatase inhibitor cocktail 3 stock 1:100.
7. Calyculin A 100 nM.
8. Bradford reagent concentrate (Bio-Rad Protein Assay; Bio-Rad Laboratories) and bovine serum albumin (BSA; Sigma or other high-grade suppliers).
9. 96-Well plates and plate reader (for protein assays).
10. 0.1 M NH_4HCO_3 (pH ~8; no pH adjustment is performed).
11. Microcentrifuge tubes, 2.0 ml (Sarstedt, part # 72.694.005).

2.5 Protein Digestion, Reduction, Alkylation, and Desalting

1. Dithiothreitol (DTT) 50.0 mM in water.
2. Iodoacetamide (IAA) 200 mM in water (*see* **Note 5**).
3. Trypsin and trypsin re-suspension buffer (Promega, modified, sequencing grade, cat. # V5111).
4. Sep Pak Plus C18 cartridges (Waters, part # Wat020515).
5. Luer-lock Syringe, 10 ml (Beckton Dickinson (BD), part # 148232A) and 21-gauge needles (BD, reorder # 305167).

6. 50 % Acetonitrile (ACN)/0.05 % formic acid (FA) (50:50 ACN:0.1 % FA in water) (ACN from Fisher or other high-grade suppliers; FA from EMD Biosciences, cat. # 11670-1 or equivalent).
7. 2.0 % ACN/0.1 % FA and 40 % ACN/0.06 % FA.

2.6 SCX-Based Separation of Peptides from the Total (Phospho)Proteome

1. SCX column: 2.1 mm × 200 mm polysulfoethyl A, 5 μm particle size, 200 Å pores (polyLC, Inc., item #202SE0502).
2. Solvent C: 5.0 % ACN/0.1 % FA in water.
3. Solvent D: 25 % ACN/0.1 % FA, containing 500 mM KCl, in water.
4. Wash solvent (40:40:20 isopropanol (IPA):ACN:H_2O).
5. Glass sample vials and caps (Waters, P/N 186001124DV) and plastic sample vials and caps (SUN-SRi Inc., part # 501 307 and part # 501 318, or equivalent substitutes).
6. Paradigm MS4 MDLC (Michrom Bioresources or suitable substitute; *see* **Note 6**).
7. Peptide macrotrap, including stainless steel macrotrap holder (Michrom).
8. Water bath sonicator (Fisher Scientific, model FS30, or suitable substitute).

2.7 Automated Desalting of SCX Fractions

1. Solvent A: 0.1 % FA in water.
2. Solvent B: 100 % ACN.

2.8 TiO_2-Based Phosphopeptide Enrichment: A Batch Method

1. Loading buffer: 65.0 % ACN/2.0 % trifluoroacetic acid (TFA, Thermo Scientific, product # 28904) saturated with glutamic acid (10 mg/ml; Sigma, part # G1251).
2. Wash buffer 1: 65.0 % ACN/0.5 % TFA.
3. Wash buffer 2: 50.0 % ACN/0.1 % TFA.
4. Elution buffer 1: 50.0 % ACN/0.3 M NH_4OH (no pH adjustment).
5. Elution buffer 2: 5.0 % ACN/0.3 M NH_4OH (no pH adjustment).
6. TiO_2 bead slurry 10.0 mg/ml in loading buffer (beads from GL Sciences, Part # 1400B500).
7. Vortex Genie 2 (Scientific Industries) or suitable substitute.

2.9 Reversed-Phase (RP) HPLC-MS/MS (LC-MS/MS)

1. Peptide captrap and PEEK captrap holder (Michrom).
2. Analytical column: 150 mm × 0.2 mm Magic C18 (3 μm particles, 200 Å pores; Michrom).
3. Solvents A and B as described above.

4. Paradigm MS2 HPLC (Michrom; *see* **Note 6**).
5. LTQ Orbitrap Velos mass spectrometer equipped with electron transfer dissociation (LTQ OT Velos ETD; Thermo Fisher Scientific; *see* **Note 6**).

2.10 MS/MS Spectral Processing Prior to Searching Against a Protein Database

1. Standard, modern desktop computer with Windows XP operating system.

2.11 Searching MS/MS Data Against a Protein Database

1. Search engine Sorcerer™-SEQUEST® on the Sorcerer Enterprise hardware/software package (SageN Research Inc., Milpitas, CA). Alternatively, if smaller quantities of data will be collected, Sorcerer™-SEQUEST® v. 2.0 can be used (*see* **Note 6**).

2.12 Post-search Processing and Further Data Analysis

1. High-speed Internet connection to the Trans-Proteomic Pipeline (TPP; Institute for Systems Biology, Seattle, WA), which provides free-access proteomic data processing tools.
2. Software algorithms, described in Subheading 3.12.

3 Methods

Regardless of the experimental conditions or design to be utilized, it is important to assess reproducibility of (phospho)protein IDs by performing at least two biological replicate samplings (i.e., from independent cultures) of each cell population (or tissue/organ type) studied and two technical replicates of each biological replicate. For human pluripotent stem cells and their derivatives, morphological and immunocytochemical validation, karyotype analysis, and mycoplasma testing should be routinely conducted during the course of culturing and experimentation, to monitor homogeneity and cellular identity and to maintain a healthy population of self-renewing cells in the absence of spontaneously differentiating progeny. For other cell types, similar measures need to be implemented. Moreover, the cells should be grown in optimal conditions and handled with care to avoid stress-activated responses (i.e., avoid temperature shock, excessive light exposure, etc.). An example of quality control measures was described recently [11]. Preliminary experiments are advisable with pluripotent stem cells, which may be laborious or expensive to culture in larger quantities required for a total (phospho)proteome analysis. Smaller scale cultures may be used to estimate the quantity of cells needed to obtain protein of sufficient yield and quality and can provide experience in specific procedures required for successful culture. Yield can be tested with standard, reliable protein assays, including but not limited to the Bradford assay. Protein quality can be tested, for example, with Western blots using antibodies against known, relevant phosphorylation sites.

Treatment and/or differentiation periods should be carefully controlled, and harvesting of cells should be done gently, promptly, and uniformly. High-quality cellular material, proper sample preparation, and careful analytical procedures are essential to minimize experimental variability and generate reliable and biologically informative datasets. It is time and labor intensive to perform a total (phospho)proteome analysis, so it is vital to perform each step accurately for the most optimal results possible at the end of the time-consuming procedure (often months).

3.1 Cell Culture

1. Culture and passage cells with MEF-CM plus FGF2 at 20 ng/ml (*see* **Note 1**) on feeder-free Geltrex-coated 6-well plates (*see* **Note 7**), to avoid contamination from feeder cells, for proteomics applications. If the starting cell population is growing on feeders, *see* **Note 8**.
2. Examine the cells under a dissecting microscope followed by inverted phase contrast viewing at 10× for evaluation of cell morphology. If any colony areas appear differentiated, mark the differentiated area with a pen on the undersurface of the dish.
3. In a sterile, laminar flow hood, tilt the plate and aspirate the media with a 2.0 ml suction pipette from the edge of the meniscus at the top of the vertical wall of the well so that adherent cells are not disturbed.
4. Directly apply suction to the marked areas to remove differentiated cells.
5. Replace the MEF-CM + FGF2 by applying a pipette to the sidewall of the well and allowing the media to gently drip into the well.
6. Examine under the dissecting microscope to verify that differentiated cells have been removed.
7. Geltrex coating of 6-well plates: Thaw Geltrex overnight at 4 °C and dilute 1:12 using DMEM/F12 (cold). Add 2 ml of the diluted Geltrex to each well of a 6-well plate, wrap in parafilm, and place at 4 °C overnight. Typically two 6-well plates of approximately 75–80 % confluence are sufficient for two technical replicates of each of the two biological replicates for the (phospho)proteomics workflow. Such numbers are typically acquired within approximately 4–6 days after passaging. Repeat **steps 2–6** until sufficient number of cells are obtained and proceed to Subheading 3.2. If sufficient cell numbers and quality are not obtained, proceed to **step 8** to passage cells. Passaging may be performed by one of several techniques commonly employed. A method of mechanical passaging is described below.
8. Scrape undifferentiated cells, avoiding marked areas, using a 2.0 ml sterile pipette in a gentle circular motion with the pipette at a 45° angle relative to the plate. Replace the lid and

swirl the plate. Examine the plate visually or under a dissecting microscope for resuspension of the cells. If approximately 80 % or more of the cells have been re-suspended, transfer the suspension, with a sterile 10 ml pipette, to a sterile, 50 ml, conical bottomed tube. If large areas of cells have not been re-suspended, repeat the scraping, and perform the transfer after satisfactory resuspension of the cells.

9. Mix cells by pipetting gently.
10. Immediately prior to plating of stem cells, ready a pre-coated Geltrex plate (*see* **Note 7**) by carefully aspirating the DMEM/F12 to avoid disturbing polymerized matrix. Apply the cell suspension to the sidewall of the well without disrupting the Matrigel. ROCK inhibitor may be added to a final concentration of 10 μM to improve cell survival after passaging. However, it is not added during routine media changes.
11. Agitate the plate thoroughly by hand to distribute the cell suspension evenly.
12. Incubate at 37 °C/5 % CO_2 (humidified). Cells begin to adhere within an hour, although they may continue to adhere overnight.
13. Repeat **steps 2–7** as necessary to acquire sufficient cell numbers.

3.2 Lysis of hESCs, hiPSCs, or Derivatives for Protein Preparation

(For each of these cell populations, two 6-well plates containing 12.0 ml total media should yield >1 mg of protein.)

1. Gently remove one plate from the incubator, place plate on wet ice, and aspirate the media, avoiding contact with the cells. After the bulk of the media is gone, hold the plate vertical for 5 s, while still aspirating, for reasonably complete yet prompt media removal.
2. Rinse each well gently with 2.0 ml of ice-cold PBS. Aspirate wells 1–3 in the same manner. Leave the PBS in wells 4–6.
3. Add 1.0 ml of ice-cold lysis buffer to well 1. Scrape cells in the first well, with the plate on ice, with a cell scraper. Transfer the lysate to well 2 with a pipetman. Scrape cells in well 2 in the same manner. Transfer the lysate to well 3. Scrape cells in well 3 in the same manner. Develop a prompt but thorough, consistent, reproducible scraping technique, and use it every time. Transfer the lysate containing cells from wells 1–3 to a microcentrifuge tube and store on ice.
4. Aspirate PBS from wells 4–6 as described above. Add 1.0 ml of ice-cold lysis buffer to well 4. Scrape cells, transfer the lysate to well 5, scrape cells, transfer the lysate to well 6, scrape cells, transfer the lysate to a microcentrifuge tube, and store on ice as described above.

5. Repeat **steps 1–4** for subsequent plates if needed.
6. Place all lysates on a rocker platform at ~1 Hz in an ice bucket for 30 min, with the tube long axes at a ~45° angle relative to the plane of the floor to thoroughly agitate the contents and facilitate protein extraction and solubilization. Alternative agitation devices to accomplish similar thorough mixing, without frothing, are acceptable.
7. Spin lysates for 20.0 min at 15,000×*g*, 2 °C. Transfer supernatants to new tubes.
8. Repeat **step** 7 to obtain the clarified lysate. Avoid the pellet, which may be soft, but otherwise maximize the yield of clarified lysate.

3.3 $(NH_4)_2SO_4$ Precipitation, Protein Storage

1. Measure the volume of the clarified lysate with a pipetman P1000 or a similar tool and add 0.598 g/ml of $(NH_4)_2SO_4$ directly into each lysate.
2. Place the lysates on a rotator wheel or platform overnight at 4 °C to precipitate total proteins from the clarified lysates (*see* **Note 9**).
3. Centrifuge the tubes at 15,000×*g*, 20 min, 2 °C. Carefully remove the supernatant and store pellets at −80 °C. This is a favorable time in the procedure to pause, because protein pellets are stable at −80 °C in our experience.

3.4 Protein Resuspension, Gel Filtration, and Quantification

1. Rinse each pellet with 1.0 ml of ice-cold Dulbecco's PBS that is 85 % saturated with $(NH_4)_2SO_4$ (608 g/l), containing 100 nM Calyculin A and 1:100 dilutions of Sigma phosphatase inhibitor cocktails 1, 2, and 3 (*see* **Note 4**). Vortex the pellets gently in the wash buffer for 1.0 min. Centrifuge at 14,000 ×*g*, 20.0 min, 4 °C and remove the supernatant. Note that protein pellets float on the dense $(NH_4)_2SO_4$-containing buffers. Keep the pelleted proteins ice-cold.
2. For protein resuspension and gel filtration, prepare 6.0 ml of 100 mM NH_4HCO_3 buffer, pH 8, containing 8 M urea, 100 nM Calyculin A, and a 1:100 dilution of Sigma phosphatase inhibitor cocktails 1, 2, and 3 (*see* **Note 4**). Gently, without introducing bubbles and especially without frothing, resuspend 1 pellet in 1.0 ml of this buffer.
3. Perform gel filtration to deplete small-molecule contaminants from the proteins using PD-10 columns. This is according to the manufacturer's protocol, and with phosphatase inhibitors present the entire time (**steps 4–10** assume processing of one sample at a time, but with experience, parallel sample processing is feasible). Before beginning, equilibrate the column with 25.0 ml of 8.0 M urea/100 mM NH_4HCO_3 (*see* **Note 10**).

4. In preparation for collecting the proteins, add 1.0 ml of the resuspension buffer to a conical bottomed, disposable, 15 ml centrifuge tube.
5. Load the entire 1.0 ml protein sample followed immediately by an additional 0.5 ml of protein resuspension buffer onto the column. Alternatively, it is acceptable to have the protein re-suspended in a total volume of 1.5 ml, and add it all to the column at this step.
6. Elute the column into the 15 ml tube from **step 4**, using 3.5 ml of 100 mM NH_4HCO_3 buffer pH 8 containing 8.0 M urea, 100 nM Calyculin A, and a 1:100 dilution of Sigma phosphatase inhibitor cocktails 1, 2, and 3 (buffer from **step 2**; *see* **Note 4**). Place the tube containing the protein suspension on ice immediately after collection.
7. Protein quantification: Perform triplicate Bradford protein assays on small aliquots of the recovered proteins to estimate protein concentration and total amount of protein recovered.
8. Protein storage: Dilute the protein suspension to 4.0 M urea with 80 % H_2O/20 % glycerol (final concentration of 10 % glycerol; *see* **Note 11**). Divide suspensions into 1.5 ml screw cap Sarstedt tubes such that the total mass of protein is 1 mg or a convenient fraction of 1 mg (e.g., 0.25 mg, 0.5 mg) and the volume is 500 μl or less, and then store at −80 °C.

3.5 Protein Digestion, Reduction, Alkylation, and Desalting

1. Thaw enough tubes of frozen protein to obtain 1 mg, and then store on wet ice.
2. Thaw one tube containing 40.0 μl (20.0 μg) of trypsin; formulations not already in the resuspension buffer are also acceptable.
3. Add 20.0 μg of trypsin to 1 mg of protein and shake overnight at 37 °C, 600 rpm.
4. Add 50.0 mM DTT stock to a final concentration of 10.0 mM. Incubate for 30.0 min at 37 °C, 600 rpm.
5. Immediately after DTT incubation, add 200 mM IAA stock, freshly prepared, to a final concentration of 20 mM and incubate for 45 min at 37 °C, 600 rpm, in the dark (*see* **Note 5**).
6. Add one volume, equaling the current volume in the tube, of 0.1 M NH_4HCO_3 pH ~8 and 20 μg more trypsin per 1 mg of protein and incubate overnight at 37 °C, 600 rpm.
7. Peptide desalting, and drying: Prepare Sep Pak Plus C18 cartridges in conjunction with a 10 ml syringe to push the liquid through the cartridge. Use a flow rate of ca. 5 ml/min, but use ca. 2 ml/min for loading the sample on the cartridge and eluting the desalted sample.
8. Wash the syringe once with 10 ml of ACN, and then wash once with 10 ml of 50 % ACN/0.05 % FA.

9. Wash the cartridge on the syringe once with 10 ml of 50 % ACN/0.05 % FA, and then wash once with 10 ml of 2.0 % ACN/0.1 % FA. The syringe and cartridge should be clean, and the cartridge is thus equilibrated under aqueous conditions to enable peptide binding to the reversed-phase media.
10. Fill the syringe with 0.5 ml of air, 1.0 ml of 2.0 % ACN/0.1 % FA and the entire digest containing tryptic peptides. Load the peptides onto the cartridge (ca. 2 ml/min flow rate). The air is to enable complete peptide loading. Once this is accomplished, minimize the amount of air flowing through the cartridge.
11. Desalt the sample twice, each time with 10.0 ml of 2.0 % ACN/0.1 % FA.
12. Fill the syringe with ca. 0.5 ml of air (for flow of all of the elution solvent through the cartridge), and elute the sample with 1.6 ml of 60 % ACN/0.1 % formic acid, flow rate ca. 2 ml/min, into a 2.0 ml Sarstedt microfuge tube.
13. Using a 21 gauge needle, poke two holes in the microfuge tube cap for the vapors to escape but to prevent sample losses due to bumping [12], and dry in a speed vac at 35 °C. Replace the cap with one lacking holes and store at −80 °C.

3.6 SCX-Based Separation of Peptides from the Total (Phospho)Proteome

1. Wash 64 plastic vials by filling with IPA, aspirate, repeat the washing with methanol, aspirate twice, and air-dry the vials. Following addition of aliquots of water, mark 32 of the vials with lines to delineate 300, 350, and 400 μl, to estimate the remaining volumes after partial dry-down of SCX fractions. Remove the water from the vials.
2. The SCX separation is done using automated Paradigm MS4 MDLC instrumentation, including an auto-sampler/fraction collector, binary–binary HPLC, and UV detector (Michrom). Xcalibur v. 2.0 software (Thermo Fisher Scientific) with custom plug-ins (Michrom) controls the instrument functions ([13]; *see* **Note 6**).
3. Prepare/clean/condition a polysulfoethyl A column with three blank SCX gradients in which a 95:5 mix of solvents C:D is injected as a blank sample in place of a sample containing peptides. A schematic diagram of the plumbing at the time of sample injection is shown in Fig. 1a. The first two blank gradients are as described previously [13]. (This SCX gradient, at a flow rate of 200.0 μl/min, is 5.0 % (solvent) D–9.0 % D from 0 to 1.0 min, 9.0–20.0 % D from 1.0 to 24.0 min, 20.0–40.0 % D from 24.0 to 34.0 min, 40.0–100.0 % D from 34.0 to 44.0 min, 100.0 % D from 44.0 to 45.0 min, 100.0–5.0 % D from 45.0 to 46.0 min, and 5.0 % D from 46.0 to 50.0 min.) The third blank SCX gradient (in which the data is saved as a baseline, and typically with an absorbance at 214 nm (A_{214}) of ca. 4 % the A_{214} of the SCX separation), at a flow rate of 200.0 μl/min,

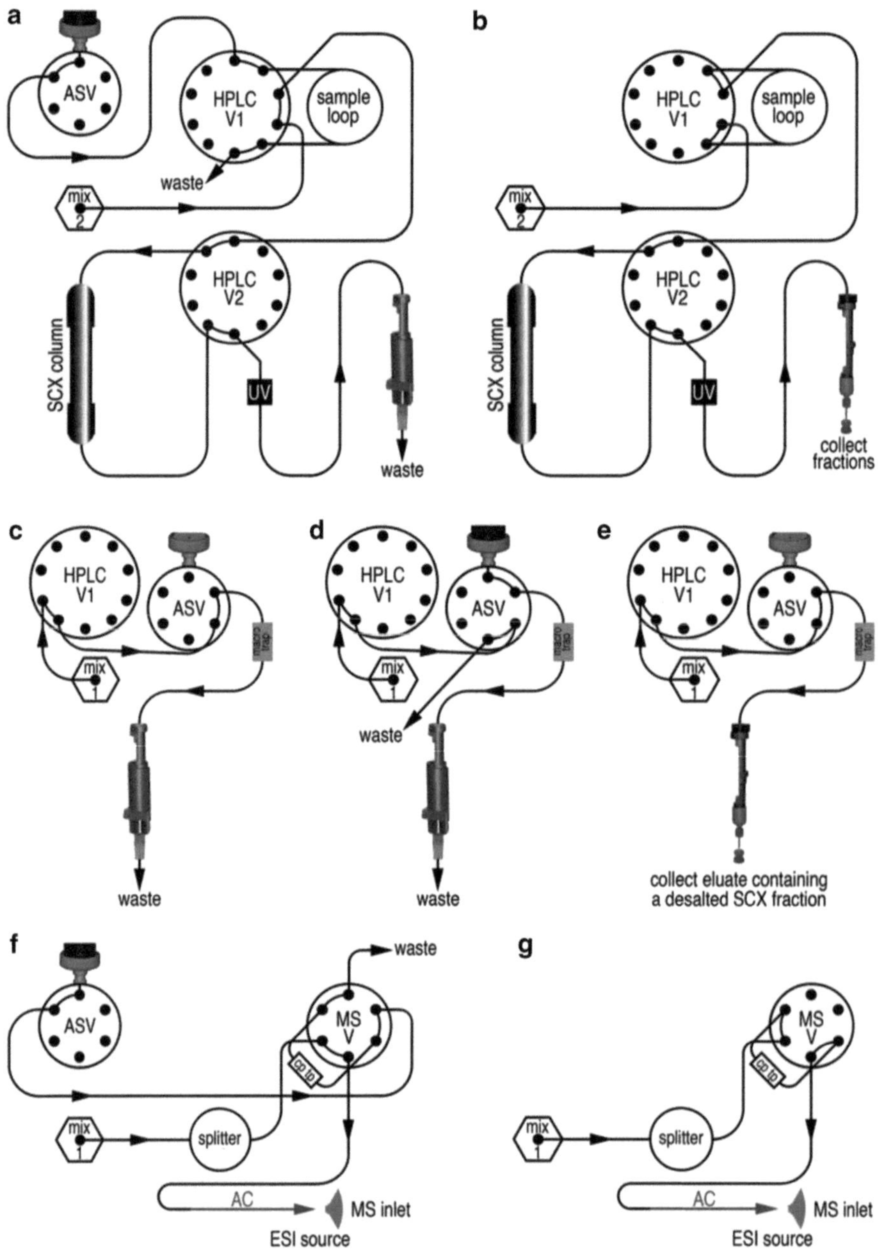

Fig. 1 Panels (**a**, **b**) are reproduced from ref. 13, and panels (**f**, **g**) are adapted from ref. 13 with permission of Elsevier Publishing. A simplified schematic diagram of the plumbing for liquid flow through the HPLC, auto-sampler, UV detector flow cell, SCX and RP columns, as well as macrotrap is shown. All tubing is 1/16′′ outer diameter (OD) PEEK, with varied inner diameters, with the exception that PEEKsil tubing (also 1/16′′ OD) is used for all sample flow paths. Tubing was obtained primarily from Michrom and also from Upchurch Scientific (Seattle, WA). (**a**) Flow paths for sample introduction and SCX column stabilization, immediately prior to SCX separation, are shown. Abbreviations: *ASV* auto-sampler valve, *HPLC V1* HPLC valve 1, *mix 2* mixer 2 (for solvents C and D); *HPLC V2* HPLC valve 2, *UV* UV detector flow cell. Two flow paths, marked by *arrows*, are active at this stage: (1) Flow of the sample, from the syringe, for sample introduction into the (100 μl) sample loop, and (2) flow from pumps C and D (not shown for clarity), through mixer 2, for stabilization of solvent flow and

is 5.0–9.0 % D from 0.0 to 2.0 min, 9.0–24.0 % D from 2.0 to 30.0 min, 24.0–44.0 % D from 30.0 to 42.0 min, 44.0–100.0 % D from 42.0 to 54.0 min, 100.0 % D from 54.0 to 57.0 min, 100.0–5.0 % D from 57.0 to 60.0 min, and 5.0 % D from 60.0 to 64.0 min (Fig. 2).

4. Starting ca. 40 min into the beginning of the blank gradients, resuspend, at room temp, the sample (desalted peptides from 1 mg of total protein, described above) in 80.0 μl of 95:5 solvent C:D. Use a resuspension method comprising maximum vortexing for 30 s, vortexing at setting 6 for 10.0 min, maximum vortexing for 30 s, sonication for 10.0 min in a water bath sonicator, and then maximum vortexing for 30 s. Effective, alternative peptide resuspension procedures are acceptable. Centrifuge at 14,000 ×*g* for 10 min at room temperature and transfer the supernatant into a glass sample vial. If there is a pellet, avoid including any of it in the sample vial to prevent fouling of the MDLC system.

Fig. 1 (continued) pressure through the SCX column prior to the SCX separation. (**b**) Flow path during the SCX separation, which no longer involves sample introduction. To enter this stage following that shown in panel (**a**), HPLC valve 1 switches, resulting in reversal of the sample flow back out of the sample loop and into the SCX column. Concurrent to the switching of valve 1, the auto-sampler automatically suspends the fraction collection tool above the vials, in order to collect eluted fractions from the SCX column. The fraction collection tool is automatically moved to a fresh vial every 2.0 min, resulting in collection of 32 × 400 μl fractions. The SCX gradient, consisting of solvents C and D, is applied, resulting in peptide separation throughout the gradient. (**c**) Flow path for solvents A and B, with the auto-sampler valve in position 1, during portions of the macrotrap wash and auto-desalting methods, when the HPLC effluent is directed through the macrotrap to the waste. The flow (400 μl/min) is unsplit, so the HPLC splitter and pumps A and B are not shown for clarity. *mix1* mixer 1. (**d**) Flow path for solvents A and B, with the auto-sampler valve in position 2, during portions of the macrotrap wash and auto-desalting methods, when the HPLC effluent is diverted to the waste without passing through the macrotrap, and reagents are injected from the syringe through the macrotrap. As in panel (**c**), the flow (400 μl/min) is unsplit. (**e**) Flow path for solvents A and B, with the auto-sampler valve in position 1, during the step-gradient portion of the auto-desalting method only, when the HPLC effluent is directed through the macrotrap and into a vial to collect the peptides from the desalted SCX fraction. As in panel (**c**), the flow (400 μl/min) is unsplit. (**f**) Flow paths for sample introduction immediately prior to reversed-phase (RP) separation. Abbreviations: *MSV* valve mounted on the mass spectrometer, *cp tp* capillary peptide trap (i.e., captrap; polymeric); and AC, C18 reversed-phase, capillary-scale (1/16″ OD) analytical column. Two flow paths, marked by *arrows*, are active at this stage: (1) Flow of the sample, from the syringe, for sample introduction onto the peptide cap trap (cp tp) followed by desalting of the captured peptides, and (2) flow from pumps A and B (not shown for clarity), through mixer 1, for stabilization of solvent flow and pressure through the C18 analytical column (AC). The high voltage (1.4 kV) is manually activated at the start of this stage using the tune page, which is part of the control software for the mass spectrometer. (**g**) Flow path during the reversed-phase separation, which no longer involves sample introduction. To enter this stage following that shown in panel (**f**), the valve mounted on the mass spectrometer switches, resulting in reversal of the solvent flow back through the peptide captrap, gradient elution of the peptides off of the peptide captrap, their separation in the analytical column, elution, and ionization in the ESI source, just prior to introduction into the mass spectrometer

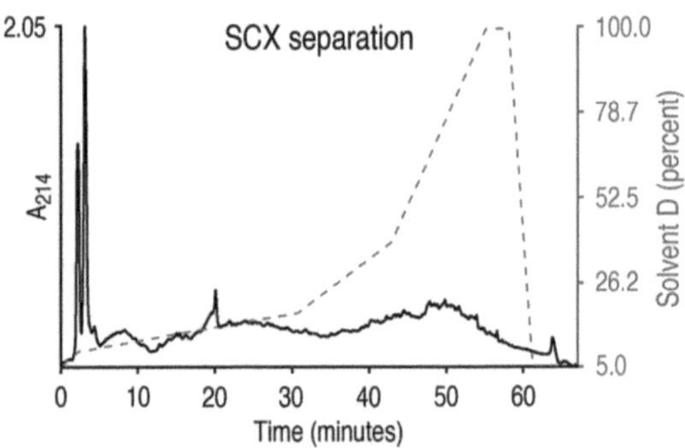

Fig. 2 SCX separation of a peptide mixture comprising the total (phospho) proteome from multipotent derivatives of pluripotent hESCs. The solid line is a typical trace of an SCX separation, in which A_{214} was recorded. Peptides derived from widely differing protein sources yield indistinguishable and consistent results. The *dashed line* indicates the HPLC gradient

5. Immediately after the last blank SCX gradient, allow the solvent to run at 200.0 μl/min for 5.0 min at 95:5 solvent C:D (flow-path shown in Fig. 1b, but with flow to the waste) to ensure that the sample loop and the column are fully equilibrated at the initial solvent condition. Initiate an SCX separation by injecting all 80 μl of the peptide suspension (Fig. 1a). In the instrument method, include 10 μl of 95:5 solvent C:D pre-sample solvent to facilitate injection of the entire sample into the 100 μl sample loop of the MDLC. The SCX gradient (Fig. 2) is the longer one described above. Throughout the SCX separation, a representative trace of A_{214} is shown (Fig. 2), and 32 × 400 μl (2.0 min) SCX fractions are collected.
6. Wash the column with one of the shorter blank SCX gradients with solvents C and D, and once more with 40:40:20 IPA:ACN:H_2O wash solvent (to clean the column and prepare it for storage at 4 °C).
7. Perform a partial dry-down of all 32 SCX fractions for 15.0 min in a speed vac (with the goal being depletion of ACN to facilitate binding of peptides to the macrotrap during automated desalting). Do NOT pipette the fractions; perform the partial dry-down in the collection vials. Cap and store the vials at 4 °C.

3.7 Automated Desalting of SCX Fractions

1. Estimate the remaining volumes of each SCX fraction by visual inspection of the vials, using the lines marked on them earlier. (Do NOT pipette the liquid containing the fractions, in order to minimize sample losses.) Set the volume in the sample cue, for the automated desalting method (below), to fill a syringe on the auto-sampler with ca. 30 μl more than the estimated remaining volume of each SCX fraction, in order to fill the syringe with the entire SCX fraction. There should be a small

volume of air in the syringe after its filling, which should not cause difficulties.

2. Prewash a peptide macrotrap (Michrom) twice using the MDLC (Fig. 1c, d). The flow rate is 400 μl/min during the entire method, initially with 98 % solvent A/2 % solvent B through the macrotrap, with the auto-sampler valve in position 1 (Fig. 1c). Filling of the syringe on the auto-sampler immediately commences with 440 ul of wash solvent at a fill speed of 9.0 μl/s. When the syringe is filled, the auto-sampler valve switches to position 2 (Fig. 1d), the HPLC flow is directed to waste without flowing through the macrotrap, and the syringe is emptied through the macrotrap at an injection rate of 5.0 μl/s. The auto-sampler valve switches back to position 1 (Fig. 1c), the HPLC flow (98:2 solvent A:B) is through the macrotrap, and the syringe is filled at a rate of 9.0 μl/s with 500 μl of 98:2 solvent A:solvent B. The auto-sampler valve switches to position 2 (Fig. 1d), and the syringe is emptied through the macrotrap at an injection rate of 5.0 μl/s. The auto-sampler valve switches to position 1 (Fig. 1c) and a step gradient is applied to the macrotrap, consisting of 2.0 % solvent B at 0.0 min, 80.0 % B at 0.06 min, 80.0 % B at 0.55 min, 2.0 % B at 0.60 min, and ending with 2.0 % B at 2.00 min.
3. Desalt each of the fractions, in the order 1–32, with the macrotrap desalting method. For the desalting method, a flow rate of 400 μl/min of 98 % solvent A/2 % solvent B is continued through the macrotrap following the previous (macrotrap-washing or auto-desalting) method, with the auto-sampler valve in position 1 (Fig. 1c). The syringe on the auto-sampler is filled with the entire SCX fraction (ca. 290–370 μl following 15.0 min in the speed vac) at a fill speed of 9.0 μl/s. The auto-sampler valve switches to position 2 (Fig. 1d) and the syringe is emptied through the macrotrap at a rate of 5.0 μl/s, resulting in binding of peptides to the macrotrap. The auto-sampler valve switches back to position 1 (Fig. 1c), the HPLC flow (98:2 solvents A:B) through the macrotrap begins desalting of the peptides, and the syringe is filled at a rate of 9.0 μl/s with 500 μl of 98:2 solvents A:B. The auto-sampler valve switches to position 2 (Fig. 1d) and the syringe is emptied through the macrotrap, to thoroughly desalt the peptides, at a rate of 5.0 μl/s. The auto-sampler valve switches to position 1 (Fig. 1e), redirecting HPLC flow through the macrotrap. Concurrently, the auto-sampler/fraction collector initiates collection of the effluent from the macrotrap, and collection continues during the following step gradient at 400 μl/min: 2.0 % (solvent) B at 0.0 min, 50.0 % B at 0.56 min, 50.0 % B at 1.05 min, 2.0 % B at 1.10 min, and ending with 2.0 % B at 2.00 min. The drop collection function of the auto-sampler/fraction collector is enabled.

After the first-, second-, and subsequently every 5 SCX fractions are desalted, the macrotrap wash method is repeated before proceeding with desalting the next SCX fraction. This macrotrap wash method could be important to prevent fouling of the plumbing.

4. Half of the eluted fraction containing each desalted SCX fraction is transferred into a 1.5 ml Sarstedt tube (resulting in two aliquots of each desalted SCX fraction, each ca. 420 μl, which enables repeat phosphopeptide enrichments of desalted SCX fractions if necessary).
5. Dry the desalted SCX fractions overnight at 35 °C in a speed vac, with holes poked in the caps of the microfuge tubes using a 21 gauge needle, two holes/cap (vapors escape through the holes, and having the caps on likely controls losses due to bumping; [12]). The next morning, put a new cap (lacking holes) on each tube and store at −80 °C.

3.8 TiO_2-Based Phosphopeptide Enrichment: A Batch Method

This protocol is modified from ref. [14]. We typically perform TiO_2 enrichments on 13 or 14 desalted SCX fractions simultaneously. In addition to elution fractions containing enriched phosphopeptides, flow-through and wash fractions are also collected. Although most detectable phosphopeptides are in TiO_2 elution fractions, the flow-through plus wash fractions contain phosphopeptides that do not bind to the TiO_2, some of which are detectable in the mass spectrometer, and they also yield a wealth of information on the total proteome.

1. A TiO_2 slurry is vortexed at setting 6.0 on a Scientific Industries Vortex Genie 2 for 10.0 min at room temp. Alternative methods to suspend the TiO_2 particles, which result in a uniform suspension, are acceptable.
2. Add 40.0 μl of the TiO_2 slurry into each sample tube, which contains a desalted and dried SCX fraction, and add 160 μl of loading buffer (*see* **Notes 12** and **13**). Vortex the tubes at setting 6.0 for 60.0 min at room temp.
3. Immediately after initiation of vortexing in **step 2**, wash plastic vials, one for each TiO_2 enrichment, with methanol and IPA as described above, for use in collecting the flow-through and wash fractions. Ensure that the vials are dry before use.
4. Centrifuge the tubes (containing TiO_2) for 60 s at RT, 14,000 ×*g*, in a microcentrifuge (Eppendorf 5415c or an equivalent) and save the supernatant (flow-through fraction) in the plastic vials from **step 3** (one vial is used for each TiO_2-based enrichment). Alternative centrifuges are acceptable. Store the vials at 4 °C. Add 200 μl of loading buffer to each sample tube, and vortex at setting 6.0 for 30.0 min at room temperature.

5. Pipette 50.0 μl of HPLC-grade water into glass vials, one vial for each phosphopeptide enrichment. Use an ultrafine-tipped sharpie to mark the vertical location, on the sides of the vials, of the meniscus of the water. Repeat three times so that volumes of 50, 100, 150, and 200 μl are marked on the sides of the vials. Discard the water from the vials.
6. Centrifuge the tubes containing the TiO_2 as in **step 4**. Pool the supernatant (wash fraction 1) in the same plastic vial (which contains the flow-through fraction from the same enrichment), and store the vials at 4 °C. Add 200.0 μl of wash buffer 1 to the sample tube, and vortex at setting 6 for 30.0 min at room temperature.
7. Centrifuge the tubes as in **step 4**. Pool the supernatants (second wash fraction) into the same plastic vials. Add 200.0 μl of wash buffer 2 to each sample tube, and vortex at setting 6.0 for 30.0 min at room temperature.
8. Centrifuge the tubes as in **step 4**. Pool the supernatants (third wash fractions) into the same plastic vials. Add 200.0 μl of elution buffer 1 to the sample tubes, and incubate at 45 °C, 1,400 rpm, for 1.0 h on a Thermomixer (Eppendorf).
9. Partially dry the flow-through plus wash fractions in a speed vac at 35 °C during the incubation listed in **step 8**.
10. Centrifuge the TiO_2-containing tubes as in **step 4**. Collect the supernatants (first elution fractions) in a glass sample vial (from **step 5**) for each elution fraction. Add 200 μl of elution buffer 2 to the sample tubes, and incubate at 45 °C, 1,400 rpm, for 1.0 h.
11. Partially dry the flow-through plus wash as well as the first elution fractions in a speed vac at 35 °C during the incubation listed in **step 10**.
12. Centrifuge the tubes as in **step 4**. Pool the supernatants (second elution fractions) in the same glass vials that contain the partially dried first elution fractions. Partially dry elution and flow-through plus wash fractions (glass vials and plastic vials, respectively) in a speed vac at 40 °C until the sample volumes are less than 50 μl. Some elution fractions resist volume reduction; for those, increase the temperature in the speed vac to 45 °C, and move the vials to the positions that contained vials that dried faster.
13. Add 100.0 μl of 95:5/solvent A:B to each TiO_2 fraction. Flow-through plus wash fraction volumes can be measured with a pipetman. Do NOT pipette/otherwise handle elution fractions to avoid losses of peptides, many of which are at very low concentrations. Estimate the elution fraction volumes on the basis of the lines on the vials (**step 5**) and store all fractions at 4 °C. Analyze the fractions by LC-MS/MS.

3.9 RP HPLC-MS/MS (LC-MS/MS)

1. Instrumentation comprises an HPLC RP column, captive spray ESI source, and an LTQ Orbitrap Velos equipped with ETD (*see* **Note 6**). Instrumentation is controlled by Xcalibur v. 2.6.0 build software (Thermo Fisher) with plug-ins (Michrom). Be aware that improved instrumentation control software could potentially be available as its development by the manufacturer progresses.
2. Prior to analysis, and especially after running previous samples, it is important to minimize sample carryover, as well as to keep the HPLC clean. To do so, HPLC packing and stationary-phase media are stripped and cleaned by injecting 100 μl of trifluoroethanol (TFE, an effective solvent for minimizing peptide carryover; [15]) twice over the polymeric cap trap (with retention characteristics similar to C8 RP media (Michrom); Fig. 1f) and running two 15-min blank gradients (Fig. 1g) over the C18 column using 40:40:20 wash solvent. The solvent A channel is transitioned into 40:40:20 wash solvent by switching over to "wash" solvent (40:40:20 wash solvent) without purging the A (0.1 % FA) solvent. After these blank gradients, the solvent A channel is switched back to 0.1 % FA; solvent B is a constant 100 % ACN. Three more identical blank gradients, with TFE injected over the captrap (Fig. 1f), are run (Fig. 1g) to further clean, condition, and transition the C18 column back into the RP solvents A and B. This routine decreases peptide carryover, may help eliminate HPLC fouling, avoid lost LC-MS/MS analyses, improve retention time consistency, and decrease the risk of damage to the HPLC plumbing and stationary-phase packing due to poor or blocked flow.
3. The LTQ OT Velos ETD is programmed to scan the precursor ions, with an m/z ratio of 300–1,500 in the Orbitrap at a resolution of 60,000 in the profile mode only. A top-20 data-dependent MS/MS method is used, and includes a decision tree to choose collision-induced dissociation (CID) or electron transfer dissociation (ETD) activation "on the fly" [16]. MS/MS scans are in the dual-pressure cell linear ion trap. Dynamic exclusion is enabled with a repeat count of 1, and repeat duration and exclusion duration for 30.0 s. The signal threshold for MS/MS is 500 counts. CID uses relative collision energy of 35.0 and an activation Q of 0.250 for 10.0 ms. ETD activation is for a maximum of 100.0 ms, and is automatically controlled by the Xcalibur instrument software.
4. Before sample loading on the Captrap, the auto-sampler injects 50.0 μl of 95:5 solvent A:B through the captrap at 2.0 μl/s (Fig. 1f). Prior to this time, the captrap and C18 column were equilibrated in 90:10 solvents A:B. It is possible that hydrophilic peptides are retained more effectively on the captrap when it is equilibrated in 95:5 (rather than 90:10) solvents A:B.

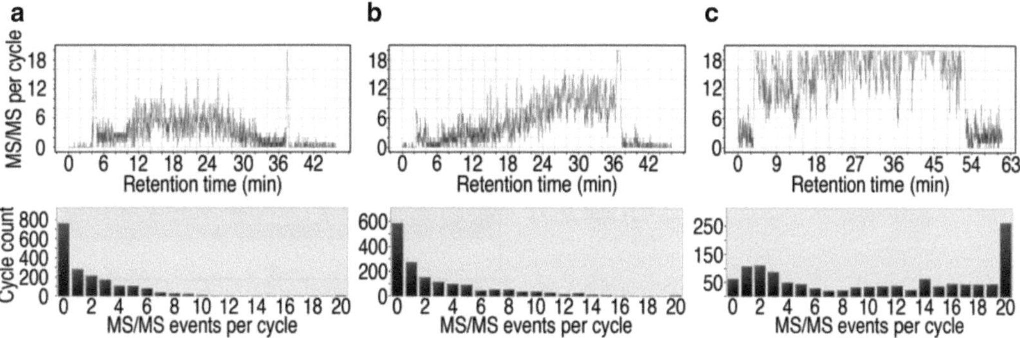

Fig. 3 Data-dependent MS/MS cycle profiling indicates the complexity of the mixtures. The number of MS/MS events per top 20 data-dependent MS/MS cycle, along with the elution profile for the MS/MS counts, is shown. (**a**) Profile for an analysis of a typical TiO_2 elution (SCX fraction 12, neural stem cells), (**b**) profile of a typical flow-through plus wash from an early SCX fraction (SCX fraction 5, neural stem cells), and (**c**) profile of a typical flow-through plus wash from a mid-gradient SCX fraction (SCX fraction 12, neural stem cells). Note the longer (45 min) gradient in (**c**)

5. The syringe is filled with a sample of the peptide mixture from a TiO_2 fraction (50 % of the elution fraction or 4 % of the flow-through plus wash fraction, each run in duplicate) and the syringe is emptied, at 0.5 μl/s into the injection port (Fig. 1f), where peptides are pushed to and captured by the captrap. Peptides on the captrap are desalted with 50 μl of 95:5/solvent A:B, which flows through the captrap at 0.5 μl/s (Fig. 1f).
6. Following sample loading and desalting, the valve on the mass spectrometer is switched to place the Captrap in line with the analytical column (Fig. 1g), and the gradient and data acquisition is activated by a "contact closure" signal from the auto-sampler.
7. For TiO_2 elution fractions, the RP gradient consists of 10.0–30.0 % solvent B from 0.0 to 30.0 min, 30.0–80.0 % B from 30.0 to 30.1 min, 80.0 % B from 30.1 to 36.0 min, 80.0–10.0 % B from 36.0 to 36.1 min, and 10.0 % B from 36.1 to 45.0 min. The same gradient is used for pooled TiO_2 flow-through plus wash fractions from SCX fractions 3–7, whereas a longer RP gradient, consisting of 10.0–30.0 % B from 0.0 to 45.0 min, 30.0–80.0 % B from 45.0 to 45.1 min, 80.0 % B from 45.1 to 51.0 min, 80.0–10.0 % B from 51.0 to 51.1 min, and 10.0 % B from 51.1 to 61.0 min, is applied to the TiO_2 flow-through plus wash fractions from SCX fractions 8–28. This is because these flow-through plus wash fractions from later in the SCX gradient are more complex than the earlier ones and the TiO_2-elution fractions (Fig. 3). Although the complexity of TiO_2-elution fractions is relatively low, as are the TiO_2 flow-through plus wash fractions early in the SCX gradient, RP gradients shorter than 30 min are not recommended by the manufacturer for the 0.2 mm × 150 mm C18 column.

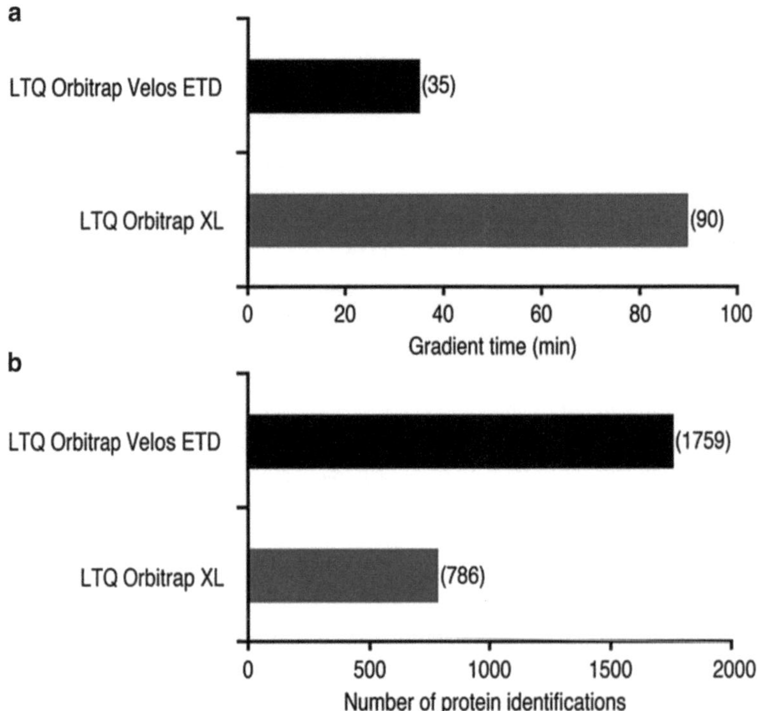

Fig. 4 An LTQ Orbitrap Velos equipped with ETD identifies more proteins in less time than an LTQ Orbitrap XL with the same sample. (**a**) Gradient times and (**b**) the number of protein identifications are shown. These typical results were from injecting 5 % of an immobilized metal affinity chromatography (IMAC) flow-through from one SCX fraction

We obtain better sensitivity with these columns than shorter (0.2 mm × 50 mm) C18 columns, which are more suitable for shorter gradients. RP gradients longer than 45 min on more complex TiO_2 flow-through plus wash fractions (Fig. 3c) can yield more protein IDs, but, with the large number of SCX-TiO_2 fractions to be run, can be time prohibitive.

8. Compared with an LTQ Orbitrap XL, which delivers very good performance, an LTQ Orbitrap Velos with ETD is capable of identifying many more proteins in a shorter time, as shown in Fig. 4. Our results are in agreement with two other recently published studies that reported improved performance with an LTQ Orbitrap Velos [17, 18].

3.10 MS/MS Spectral Processing Prior to Searching Against a Protein Database

Although high-quality results are obtained from searches using raw data with the parameters specified in Subheading 3.11 below, we have found that preprocessing ETD spectra results in improvements in the number and quality of protein identifications (*see* **Note 14**).

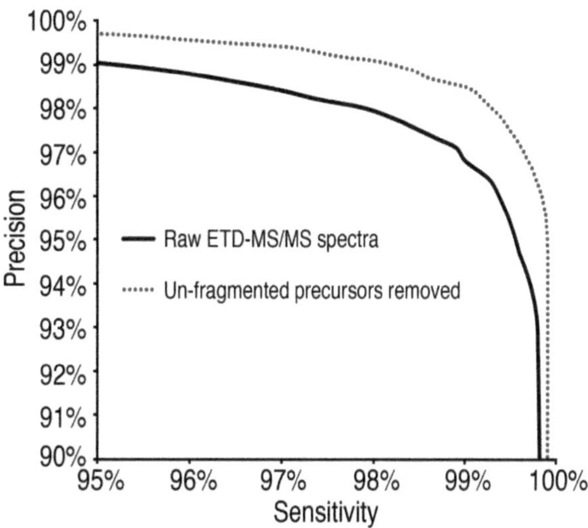

Fig. 5 Removal of signal from un-fragmented precursor ions from ETD-MS/MS spectra resulted in improved peptide identification compared to searches of the raw data. Raw data (*solid black curve*) was then processed by removing un-fragmented and charge-reduced precursor ions (*dotted curve*). Precision and sensitivity estimates were provided by PeptideProphet (TPP)

1. Convert .RAW files to .mzXML files by using the ReAdw.exe program (http://tools.proteomecenter.org/wiki/index.php?title=Software:ReAdW).
   # Command-line: ReAdw.exe --mzXML -c *.RAW.
2. Run MzXML2Search.exe (http://tools.proteomecenter.org/wiki/index.php?titleSoftware:MzXML2Search) to generate .dta files.

   # Command-line: MzXML2search.exe -B400 -T6000 *.mzXML.
3. Charge-reduced precursor ions receive one or more electrons during the ETD reaction but are not dissociated. By using an in-house script, available from the authors, precursor ions, charge-reduced precursor ions, and neutral loss ions present in the ETD spectra are removed; these ions were described previously [19] (*see* **Note 14**). For un-fragmented precursor and charge-reduced precursor ions, all peaks within ±2 m/z units of the predicted m/z of these ions are removed. Neutral loss peaks are all peaks with a mass ≤60 Da/z below that of un-fragmented precursors or un-fragmented, charge-reduced precursors [19].
4. Removing un-fragmented precursor ions (described above) from MS/MS spectra improves the precision and sensitivity of identification, as shown in Fig. 5. Removal of neutral loss products and un-fragmented precursor ions resulted in a curve that was essentially the same as the curve shown in Fig. 5, in which only un-fragmented precursors were removed (data not shown).

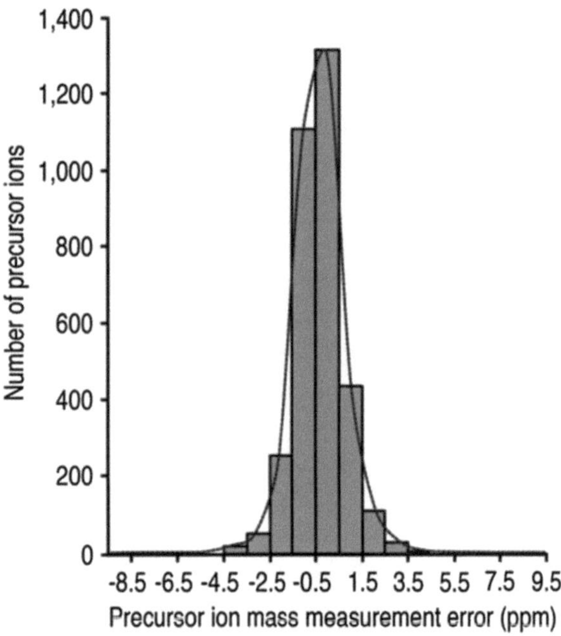

Fig. 6 The mass accuracy distribution of precursor ions, detected in the Orbitrap of an LTQ Orbitrap Velos with ETD, is presented in bins of 1.0 ppm and includes the curve outlining the precise distribution. The 99 % confidence interval of the distribution is −3.1375–2.4572 ppm, with a mean mass accuracy of −0.3402 ppm

3.11 Searching MS/MS Data Against a Protein Database

The search engine Sorcerer™-SEQUEST® on the Sorcerer Enterprise hardware/software package is used.

1. Precursor ion mass tolerance of 5 parts per million (ppm) and fragment mass type, monoisotopic, is specified. Searching data with mass tolerance of relatively optimal stringency yields an increased number of high-confidence peptide IDs (we specify a false discovery rate (FDR) of 0.005–0.009), especially for low-abundance peptides, such as phosphopeptides. Typically, the mass error of precursor ions measured with an LTQ Obitrap is less than 5 ppm. A typical mass accuracy distribution, with external mass calibration according to the recommendations of the mass spectrometer manufacturer, is shown (Fig. 6).
2. Product ion mass tolerance, for MS/MS scans performed in the linear ion trap, is 0.5 atomic mass units (amu).
3. The following differential modifications are specified: Phosphorylation (+79.966331 amu) on Ser, Thr, and Tyr, and oxidation (+15.99492 amu) on Met. For ETD spectra, modifications of N-termini (b- to c-ions, +17.02655 amu) and C-termini (y- to z-radical ions, −16.018724 amu; Versasearch Script, SageN) are included to account for the higher occurrence of c- and z-type, rather than b- and y-type product ions [20, 21]. The mass of c-ions is equal to the mass of b-ions with

a +17.02655 amu modification on N-termini, and the mass of z-ions is equal to that of y-ions with a −16.018724 amu modification on C-termini. Therefore, these terminal modifications allow the SEQUEST algorithm to effectively search CID and ETD MS/MS spectra together.

4. Static modification, carboxyamidomethylation (+57.02146 amu) on Cys.
5. Isotopic check using mass shift of 1.003355 amu.
6. The database is from the international protein index (IPI). For protein samples from human pluirpotent cells, we currently use the ipi.HUMAN.v.3.73 (June 2010), with semi-tryptic specificity. The IPI databases are currently being updated, but IPI is scheduled for closure at a time not currently defined. Personnel at IPI recommend UniProtKB as the reference database for mass spectrometry in the future.
7. Four, six, or eight MS/MS data files are searched together, from duplicate LC-MS/MS analyses of the TiO_2 elution and flow-through plus wash fractions from 1 SCX fraction or 2 back-to-back SCX fractions. MS/MS data files from SCX fractions 2 and 3, 4 and 5, 6 and 7, as well as 27 and 28 are searched together, and MS/MS data files from single SCX fractions are searched together for the rest of the data.

3.12 Post-search Processing and Further Data Analysis

1. Search results are filtered at a stringent protein FDR of 0.005–0.009 using ProteinProphet (TPP).
2. We use procedures to recover lower scoring but accurate MS/MS spectra matched to highly confident phosphopeptide IDs, similar to results recently described [22]. These procedures are being described in detail [23].
3. The relative abundance of proteins that are identified is examined using spectral counts (termed "total" in the ProteinProphet XML viewer, and "*n*-instances" when the results are exported from ProteinProphet to Microsoft Excel). Normalized spectral abundance factor (NSAF)-based spectral counting is used [24]. The use of statistical analysis tools is currently being explored. In addition, in a recent comparative study using pluripotent and multipotent stem cells, we identified thousands of known and unknown proteins as well as novel phosphorylation sites defining different proteomic signatures from stem cells (Singec et al., in preparation).
4. Protein identifications are used as input for gene ontology (GO) terms, Kyoto Encyclopedia of Genes and Genomes (KEGG) pathway analyses, and GeneGo Metacore Pathway Analyses. We have also used Ingenuity Pathway Analyses, and are currently exploring the use of additional pathway analysis tools. Finally, biological validation of proteomics-based

pathway predictions will be performed, similar to previous experiments [11], but in greater depth and with more advanced tools.

4 Notes

1. To prepare MEF-CM, plate fibroblasts in tissue culture flasks and culture in DMEM/F12 + GlutaMax containing 10 % fetal bovine serum (FBS) + penicillin/streptomycin (Pen/Strep). 24 h prior to harvesting of MEF-CM, aspirate the media and replace with 20 % Knockout serum replacement (Invitrogen/Gibco, cat. # 10828028) in DMEM/F12 + GlutMAX (Gibco 10565) + 1X ß-mercaptoethanol (BME; also called 2-mercaptoethanol) (Gibco 21985) + Pen/Strep. On the following day, collect media, add Pen/Strep again and, optionally, Normocin. Pipette media to a sterile filter. Media can be transferred to 50 ml conical tubes and stored at −20 °C. Prior to use, add FGF2 to a final concentration of 20 ng/ml filtered media.
2. Stock solutions are prepared with HPLC-grade water, and all water used is HPLC grade (Sigma Chromasolve Plus). Lysis buffer stocks are filtered with a 0.22 μm filter, and are stable at room temperature for months. (HPLC solvents are NOT filtered.) The initial portion of the lysis buffer (Subheading 2.2) is prepared starting with ca. 50 % of the final volume of HPLC-grade water, the day before cell lysis (or it may be stored at 4 °C for up to 1 month), and is chilled to 4 °C overnight prior to use. Preparation of the lysis buffer is completed the next day, on a stir plate surrounded by an ice bath, immediately before the cell lysis commences.
3. For the sake of accuracy, patience and care must be exercised when pipetting viscous solutions (glycerol and NP-40). When filling pipettes, wait for the solution to stop flowing upward, eject slowly into the incipient lysis buffer, and rinse all of the material out of the pipette or pipette tip when preparing the lysis buffer, using the incipient lysis buffer to rinse.
4. Protease and phosphatase inhibitors are toxic, so handle with care! We have been using phosphatase inhibitor cocktail 1, but its availability has become uncertain. Although there may be enough phosphatase inhibitors in the phosphatase-containing buffers without it, we have also recently begun to add another phosphatase inhibitor cocktail from Thermo Fisher (Pierce) Scientific according to the manufacturer's instructions: http://www.thermoscientific.com/wps/portal/ts/products/

detail?navigationId=L11796&categoryId=92453&productId=11954764.

5. IAA is sensitive to degradation in light, so its exposure to light should be minimized. The powder should be stored in a desiccator, in the dark, at 4 °C.
6. We have had success with the instrumentation listed, but the MDLC-MS/MS methods can be adapted to other instrumentation that is available to the reader.
7. Prior to culturing the cells, prepare Matrigel-coated 6-well culture plates by diluting Matrigel 1:12 in cold (4 °C) DMEM/F12 and adding 1.0 ml into each well of 6-well plates, 24 h prior to use of the plates. Optional: Matrigel may be thawed on ice overnight, and pipettes as well as plates may be prechilled prior to plating Matrigel.
8. If pluripotent cells have been previously maintained on feeders, passage them to feeder-free conditions and passage several additional times to ensure that feeder cells are absent from the culture.
9. If any of the $(NH_4)_2SO_4$ remains undissolved after the overnight agitation, pipette off the cloudy liquid phase, if it is cloudy rather than clear, and place it in a fresh centrifuge tube, away from the $(NH_4)_2SO_4$ crystals. If precipitated proteins are bound to the sides of the tube and the liquid phase is cloudy, leave the proteins in the same tube and physically remove the crystals with extremely clean, stainless steel tweezers or some other device that will not introduce any contamination. Washing stainless steel tweezers with methanol gets them extremely clean. If the amount of undissolved $(NH_4)_2SO_4$ is substantial, precipitated proteins are bound to the wall of the tube, and the liquid phase is clear, pipette the liquid phase to a fresh tube, and then use HPLC-grade water to gently dissolve the $(NH_4)_2SO_4$ without contacting the precipitated protein. Use caution to avoid excessive losses of protein due to precipitated proteins binding to inner surfaces of pipettes, etc. The cloudy phase should be transferred back into the original tube once the $(NH_4)_2SO_4$ crystals have been successfully removed. If the liquid was clear and the precipitated proteins remained bound to the sides of the tube during removal of $(NH_4)_2SO_4$ crystals, the clear liquid can be discarded.
10. Prepare this solution for gel filtration by starting with 50 % of the final volume of HPLC-grade water. It may be necessary to add small quantities of water for the urea to dissolve, but take care to keep the volume low enough to add the rest of the components without exceeding 25.0 ml. Bring the solution to the final volume with HPLC-grade water after all other components have been added.

11. This buffer works well for protein storage at −80 °C, and for reduction/alkylation/digestion with trypsin. Failure to dilute the protein suspension as directed can result in irreversible aggregation following thawing and/or poor digestion by trypsin.
12. The TiO_2 beads rapidly settle out of the slurry, so it is necessary to vigorously pipette it up and down while dispensing to tubes, containing desalted SCX fractions, to distribute consistent quantities of TiO_2.
13. The peptide-to-TiO_2 ratio is an important factor for phosphopeptide enrichment [25], so the amount of TiO_2 slurry used in the experiment should be optimized if the quantity of peptides to be enriched differs from that stated in this protocol.
14. In contrast with typical activation by CID, abundant, unfragmented precursor ions remain following activation by ETD, as do charge-reduced precursor ions, whose presence often results in lower scoring peptide-spectrum matches, since refinements in database searching algorithms for searching ETD spectra have been in progress for a shorter time than for CID spectra [19]. Therefore, we recommend processing ETD MS/MS spectra, to remove those un-fragmented ions, for improved database searches.

Acknowledgments

We thank Anjum Khan and Dante Bencivengo of Thermo Fisher Scientific; Peter Kent, Lori Ann Upton, David Mintline, and Kerry Nugent of Michrom Bioresources for support with instrumentation; David Chiang and Patrick Chu from SageN and Kutbuddin Doctor from the Sanford-Burnham Medical Research Institute (SBMRI) for bioinformatics support; as well as Khatereh Motamedchaboki and Wenhong Zhu from SBMRI for help with some of the experiments. Support was provided by the SBMRI, NCI Cancer Center Support Grant 5 P30 CA30199-28, The La Jolla Interdisciplinary Neuroscience Center Cores Grant 5 P30 NS057096 from NINDS, and RC2 MH090011 from NIMH.

References

1. Takahashi K, Tanabe K, Ohnuki M, Narita M, Ichisaka T, Tomoda K, Yamanaka S (2007) Induction of pluripotent stem cells from adult human fibroblasts by defined factors. Cell 131:861–872
2. Thomson JA, Itskovitz-Eldor J, Shapiro SS, Waknitz MA, Swiergiel JJ, Marshall VS, Jones JM (1998) Embryonic stem cell lines derived from human blastocysts. Science 282: 1145–1147
3. Singec I, Jandial R, Crain A, Nikkhah G, Snyder EY (2007) The leading edge of stem cell therapeutics. Annu Rev Med 58:313–328
4. Brandenberger R, Wei H, Zhang S, Lei S, Murage J, Fisk GJ, Li Y, Xu C, Fang R, Guegler K, Rao MS, Mandalam R, Lebkowski J,

Stanton LW (2004) Transcriptome characterization elucidates signaling networks that control human ES cell growth and differentiation. Nat Biotechnol 22:707–716
5. Ramalho-Santos M, Yoon S, Matsuzaki Y, Mulligan RC, Melton DA (2002) "Stemness": transcriptional profiling of embryonic and adult stem cells. Science 298:597–600
6. Ideker T, Thorsson V, Ranish JA, Christmas R, Buhler J, Eng JK, Bumgarner R, Goodlett DR, Aebersold R, Hood L (2001) Integrated genomic and proteomic analyses of a systematically perturbed metabolic network. Science 292:929–934
7. Blume-Jensen P, Hunter T (2001) Oncogenic kinase signalling. Nature 411:355–365
8. Hunter T (2009) Tyrosine phosphorylation: thirty years and counting. Curr Opin Cell Biol 21:140–146
9. Schlessinger J (2000) Cell signaling by receptor tyrosine kinases. Cell 103:211–225
10. Aebersold R, Mann M (2003) Mass spectrometry-based proteomics. Nature 422:198–207
11. Brill LM, Xiong W, Lee KB, Ficarro SB, Crain A, Xu Y, Terskikh A, Snyder EY, Ding S (2009) Phosphoproteomic analysis of human embryonic stem cells. Cell Stem Cell 5:204–213
12. Brill LM, Salomon AR, Ficarro SB, Mukherji M, Stettler-Gill M, Peters EC (2004) Robust phosphoproteomic profiling of tyrosine phosphorylation sites from human T cells using immobilized metal affinity chromatography and tandem mass spectrometry. Anal Chem 76:2763–2772
13. Brill LM, Motamedchaboki K, Wu S, Wolf DA (2009) Comprehensive proteomic analysis of Schizosaccharomyces pombe by two-dimensional HPLC-tandem mass spectrometry. Methods 48:311–319
14. Hou J, Cui Z, Xie Z, Xue P, Wu P, Chen X, Li J, Cai T, Yang F (2010) Phosphoproteome analysis of rat L6 myotubes using reversed-phase C18 prefractionation and titanium dioxide enrichment. J Proteome Res 9:777–788
15. Mitulovic G, Stingl C, Steinmacher I, Hudecz O, Hutchins JR, Peters JM, Mechtler K (2009) Preventing carryover of peptides and proteins in nano LC-MS separations. Anal Chem 81:5955–5960
16. Swaney DL, McAlister GC, Coon JJ (2008) Decision tree-driven tandem mass spectrometry for shotgun proteomics. Nat Methods 5:959–964
17. McAlister GC, Phanstiel D, Wenger CD, Lee MV, Coon JJ (2010) Analysis of tandem mass spectra by FTMS for improved large-scale proteomics with superior protein quantification. Anal Chem 82:316–322
18. Olsen JV, Schwartz JC, Griep-Raming J, Nielsen ML, Damoc E, Denisov E, Lange O, Remes P, Taylor D, Splendore M, Wouters ER, Senko M, Makarov A, Mann M, Horning S (2009) A dual pressure linear ion trap Orbitrap instrument with very high sequencing speed. Mol Cell Proteomics 8:2759–2769
19. Good DM, Wenger CD, McAlister GC, Bai DL, Hunt DF, Coon JJ (2009) Post-acquisition ETD spectral processing for increased peptide identifications. J Am Soc Mass Spectrom 20:1435–1440
20. Chalkley RJ, Medzihradszky KF, Lynn AJ, Baker PR, Burlingame AL (2010) Statistical analysis of peptide electron transfer dissociation fragmentation mass spectrometry. Anal Chem 82:579–584
21. Syka JE, Coon JJ, Schroeder MJ, Shabanowitz J, Hunt DF (2004) Peptide and protein sequence analysis by electron transfer dissociation mass spectrometry. Proc Natl Acad Sci U S A 101:9528–9533
22. Zhou JY, Schepmoes AA, Zhang X, Moore RJ, Monroe ME, Lee JH, Camp DG, Smith RD, Qian WJ (2010) Improved LC-MS/MS spectral counting statistics by recovering low-scoring spectra matched to confidently identified peptide sequences. J Proteome Res 9:5698–5704
23. Singec I, Hoo J, Crain AM, Tobe BTD, Talantova M, Doctor KS et al. Deep Phosphoproteomic Profiling of Human Pluripotency and neural lineage commitment (In Preparation)
24. Zhang Y, Wen Z, Washburn MP, Florens L (2010) Refinements to label free proteome quantitation: how to deal with peptides shared by multiple proteins. Anal Chem 82:2272–2281
25. Li QR, Ning ZB, Tang JS, Nie S, Zeng R (2009) Effect of peptide-to-TiO_2 beads ratio on phosphopeptide enrichment selectivity. J Proteome Res 8:5375–5381

Chapter 13

Transcriptional Regulatory Mechanisms That Govern Embryonic Stem Cell Fate

Satyabrata Das and Dana Levasseur

Abstract

Embryonic stem cells (ESCs) are defined by their simultaneous capacity for limitless self-renewal and the ability to specify cells borne of all germ layers. The regulation of ESC pluripotency is governed by a set of core transcription factors that regulate transcription by interfacing with nuclear proteins that include the RNA polymerase II core transcriptional machinery, histone modification enzymes, and chromatin remodeling protein complexes. The growing adoption of systems biological approaches used in stem cell biology over last few years has contributed significantly to our understanding of pluripotency. Multilayered approaches coupling transcriptome profiling and proteomics (Nanog-, Oct4-, and Sox2-centered protein interaction networks or "interactomes") with transcription factor chromatin occupancy and epigenetic footprint measurements have enabled a more comprehensive understanding of ESC pluripotency and self-renewal. Together with the genetic and biochemical characterization of promising pluripotency modifying proteins, these systems biological approaches will continue to clarify the molecular underpinnings of the ESC state. This will most certainly contribute to the improvement of current methodologies for the derivation of pluripotent cells from adult tissues.

Key words Pluripotency, Self-renewal, Chromatin immunoprecipitation, Proteomics, Interactome, RNA polymerase II, Mediator, Micro RNA

1 Introduction

Embryonic stem cells (ESCs) have unlimited self-renewal and can specify cell types comprising all germ layers, a unique biological property referred to as pluripotency. The genetic mechanism that governs these characteristics is not well understood but critically informs our understanding of early developmental biology, cell fate specification, somatic cell reprogramming, and pluripotent cell regenerative medicine. The ability to characterize the genetic underpinnings of pluripotency arrived with the derivation of ESCs from the early murine embryo [1]. The availability of ESC lines and the development of homologous recombination approaches in

Nicholas Zavazava (ed.), *Embryonic Stem Cell Immunobiology: Methods and Protocols*, Methods in Molecular Biology, vol. 1029, DOI 10.1007/978-1-62703-478-4_13, © Springer Science+Business Media New York 2013

mammalian cells enabled complex genetic analyses to be performed for the first time in ESC and mice [2, 3].

The characterization of defined genetic perturbations in ESCs has contributed invaluably to our understanding of the transcriptional machinery that controls the ESC state. This culminated with the recent demonstration that differentiated adult tissues could be converted or "reprogrammed" into pluripotent cells that resemble ESCs morphologically and functionally, and exhibit a genetic and epigenetic signature that is nearly identical [4–7]. In the last few years RNA interference screens [8], transcriptome profiling [9], protein–DNA occupancy [10, 11], and protein–protein interaction [12] studies have enabled a broader picture of ESC regulation to emerge and have significantly expanded the list of participants required for pluripotency. The transcriptional regulatory mechanisms described here have primarily been gleaned from studies in murine ESCs. However, the substantial genetic overlap shared between ES and induced pluripotent stem (iPS) cell populations suggests that a better understanding of ESC function will better inform future efforts to efficiently coax murine and human iPS cells into hematopoietic lineages.

2 Core Pluripotency Factors: Oct4, Sox2, and Nanog

Oct4, a homeodomain-containing transcription factor, was the first major pluripotency factor to be discovered. Required for early development [13] and self-renewal/pluripotency of ESCs [14], Oct4 is highly regulated with gene dosage levels tolerated only within a narrow threshold [14]. Sox2 was the next intensively studied pluriopotency factor and was found to be essential for early embryonic development [15]. Oct4 and Sox2 heterodimerize and bind many developmentally regulated transcription factors. As mentioned above, an exhaustive screen for regulators of pluripotency revealed that these two factors were essential for the derivation of iPS cells [6], and in some cases are sufficient [16] if other critical genetic modifiers are expressed at the required threshold levels. Interestingly, it appears that one role of Sox2 is to positively regulate Oct4. Sox2 downregulates repressors of Oct4 and upregulates nuclear factors that activate Oct4 expression [17].

The third member of the core pluripotency triad was discovered shortly after using a screen for genes that could rescue ESCs subjected to leukemia inhibitory factor (LIF) withdrawal [18]. Nanog was the first nuclear factor identified that was capable of this feat and was soon dubbed one of the "master regulators" of pluripotency. As expected, disruption of Nanog function resulted in early failure of the developing embryo and the compromise of ESC self-renewal and pluripotency [8, 19].

Since depletion of any of the core pluripotency factors is not tolerated in ESCs, Smith and colleagues devised a transgenic system to enable controlled depletion of Oct4 [14]. The endogenous alleles of this cell line have been disrupted by homologous recombination and the cells harbor a doxycycline-inducible transgene. Upon administration of doxycycline, downregulation of transgenic protein occurs and the ESCs differentiate into a trophectoderm-like cell type. Transcriptional profiling in these cells has revealed an expanded transcriptional network controlled by Oct4 [9]. However, it has become clear that the ESC state is controlled by a larger network of transcription factors that act combinatorially to bind and regulate promoters of developmental target genes. The context of multiple factor binding at a particular promoter is essential for determining the contribution of a protein to ESC function.

3 Pluripotency Factor Binding: Combinatorial Control of ESCs

Chromatin immunoprecipitation (ChIP) has enabled the widespread determination of transcription factor occupancy at promoters. Several groups have recently used ChIP coupled with microarray technology (ChIP-chip) or high-throughput sequencing (ChIP-Seq) to map DNA binding locations of the core and accessory pluripotency factors on a genome-wide scale [10, 11]. This is a powerful approach to characterize which gene targets may be vital for the maintenance of self-renewal or pluripotency. Not surprisingly, binding was often located at promoters and regulatory elements of genes that are essential for developmental processes [10, 11]. Using gene ontology associations, there was a significant correlation with the number of core pluripotency factors bound at a particular promoter and the likelihood that its cognate gene was involved in developmental functions [11]. Higher pluripotency factor occupancy was highly predictive of a developmentally regulated transcription factor. Conversely, if a promoter was bound by only one factor, there was a much higher likelihood that its function served a housekeeping role. A core pluripotency "signature" was revealed that implicated Oct4, Sox2, Nanog, KLF4, Naccl, and Nr0b1 (aka Dax1) as having a higher propensity for binding many essential developmental regulatory genes.

We curated and mapped global coordinates of protein occupancy data that had been deposited at the Gene Expression Omnibus (GEO) [10, 11, 20]. Binding of Nr5a2 was also included since this protein is an activator of Oct4 and can compensate for Oct4 function in gene cocktails used to produce iPS cells [21]. Like the core pluripotency factors, Nr5a2 is essential for specification of the developing embryo and its genetic ablation results in lethality shortly after implantation [22]. Sall4 is required for ESC self-renewal [12] and early development, so we mapped occupancy of the two alternate splice isoforms of this nuclear protein as well [23].

Table 1
Comparison of gene targets employed in global chromatin occupancy studies

Gene	Chen et al.[a]	Kim et al.[b]	Marson et al.[c]	Hu et al.[d]	Rao et al.[e]	Heng et al.[f]
Oct4	Y	Y	Y			
Sox2	Y	Y	Y			
Nanog	Y	Y	Y			
Nr0bl		Y				
Naccl		Y				
Esrrb	Y					
Klf4	Y	Y				
Zfp42		Y				
Nr5a2						Y
Zfp281		Y				
Sall4a					Y	
Sall4b					Y	
Trim28				Y		
Cnot3				Y		
Smadl	Y					
Stat3	Y					
Tcf3			Y			
c-Myc	Y	Y				
n-Myc	Y					
Ctcf	Y					
E2F1	Y					
p300	Y					
Suz12	Y					
Tcfcp2l1	Y					
Zfx	Y					

[a]ChIP-Seq data from ref. 9
[b]ChIP-chip data from ref. 10
[c]ChIP-Seq data from ref. 42
[d]ChIP-chip data from ref. 24
[e]ChIP-chip data from ref. 22
[f]ChIP-Seq data from ref. 20

Using global protein location coordinates from the 25 factors listed in Table 1, we analyzed the genome of ESCs for binding patterns at transcription factors that are known transcriptional regulators of ESC self-renewal or are essential for hematopoietic stem and

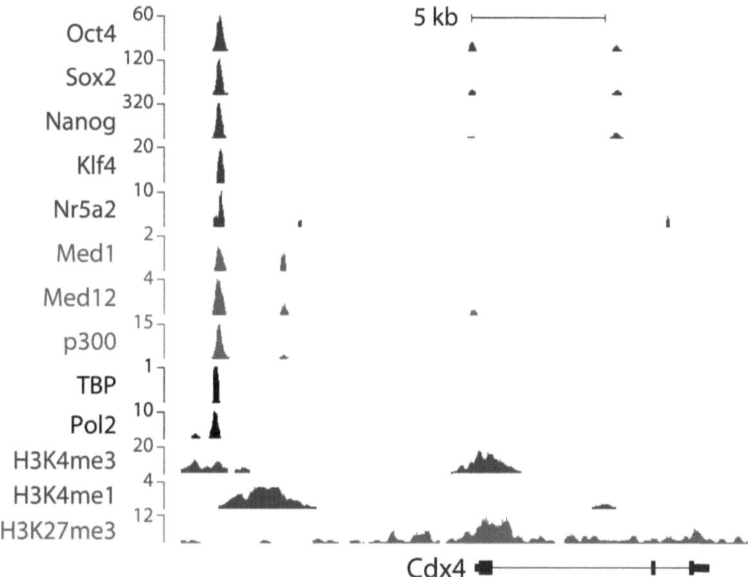

Fig. 1 Combinatorial binding of pluripotency factors at a regulator of hematopoietic specification. Genomic context of the Cdx4 gene showing pluripotency and RNA polymerase II (Pol2) chromatin occupancy. ChIP-Seq datasets for Nanog, Oct4, Sox2, Klf4, Nr5a2, Mediator components (Med1 and Med12), transcription co-regulators (TBP, p300), Pol2, and histone H3 methylation marks (H3K4me1, H3K4me3, and H3K27me3) were downloaded from the Gene Expression Omnibus (GEO, http://www.ncbi.nlm.nih.gov/geo/, accession numbers: GSE11172, GSE11431, GSE19019, GSE24164, GSE11724, and GSE22562), reads were aligned to the mouse reference (mm9) genome using Bowtie, and peaks were called with MACS to determine regions of statistically significant occupancy across the genome. The core pluripotency factors Oct4 and Sox2 and Nanog (OSN) are modestly enriched at the transcriptional start site (TSS) and downstream of the TSS. Note that depiction of Nanog marking at the TSS is obscured due to the robust occupancy at the upstream region. An alternate upstream regulatory element is marked by significant occupancy of core (OSN) and accessory (Klf4, Nr5a2) pluripotency transcription factors, and this is overlaid upon regions bound by the basal transcriptional apparatus (Pol2, Tbp) and coactivators (p300, Med1, Med12)

progenitor cells (HSC). Cdx4 is a homeobox gene that can amplify ESC-derived hematopoietic progenitors when expression is induced in transgenic ESCs [24]. We discovered that Cdx4 is significantly occupied by Oct4, Sox2, Nanog, and other critical regulators of pluripotency (Fig. 1). Although the murine and human genomes have been sequenced at significant depth for some time, annotation of protein-coding genes in both species is still incomplete. Consequently, expressed sequence tags (EST) and employment of rapid amplification of cDNA end (RACE) methodologies continue to reveal many novel alternative splice products and promoters.

Our occupancy studies suggest that transcription factor binding is also a useful method to identify putative alternative promoters and transcriptional start sites since a region located nearby the annotated Cdx4 start site is bound by five pluripotency factors and contains a transcriptional and epigenetic signature that suggests that it may be transcriptionally active in ESCs (Fig. 1). The annotated transcriptional start site contains a bivalent histone signature (discussed below) that indicates that this transcript is likely expressed at low levels or repressed in ESCs. This raises the possibility that Cdx4 may have an unanticipated role in the regulation of ESC pluripotency or self-renewal that differs from its documented HSC-promoting function [24]. RACE and transgenic expression analyses will have to be employed to verify the existence of an alternative promoter and/or splice product in HSC and ESCs. Together with the analysis of histone enrichment and RNA polymerase II (Pol II) occupancy (discussed below), identifying pluripotency factor binding to stem cell gene targets is a powerful tool for gene annotation and revealing novel DNA-binding footprints that may regulate gene expression. This provides an invaluable approach for revealing novel DNA-binding footprints that may regulate gene expression.

4 The Pluripotency Protein Network

The advent of improved proteomic methodologies has enabled the efficient coupling of protein immunoprecipitation with mass spectrometry measurement. We employed a biotinylatable peptide tagging and mass spectrometric approach to demonstrate that Nanog and Oct4 associate with numerous other critical factors that form a tight protein interaction network apparently dedicated to pluripotency. These studies allowed us to assemble an interactome network based on protein–protein interactions [12] that significantly expanded the known nuclear participants responsible for the ESC state [8, 11, 12, 25–27]. This network revealed a preponderance of proteins that are essential for early embryonic development and/or ESC self-renewal and afforded the first interactome in a pluripotent stem cell population. Similar approaches have yielded Oct4-centered interactomes that share much of the same pluripotency architecture and also reveal other regulatory factors that define ESC function [28, 29]. Collectively, the studies show a complex interwoven web of interactions between the pluripotency factors Nanog, Oct4, Sox2, Sall4, Nacc1, Nr0b1 (aka Dax1), Esrrb, TRIM28, and Zfp42 (aka Rex1). An additional notable unifying characteristic of all three interactomes is the significant presence of multiple repression complexes. Many components of the SWI/SNF chromatin remodeling, histone deacetylase and nucleosomal remodeling (NURD), and polycomb group (PcG) repression complexes are well represented within the networks, signifying that a major function of ESCs is to keep the majority of developmental genes repressed.

Using a multilayered systems biology approach, Lemischka and colleagues recently coupled transcriptome profiling and proteomics with RNA Pol II chromatin occupancy and the transcriptionally permissive histone H3 lysine 9 and 14 acetylation (H3K9/14ac) mark to determine the effect on the transcriptome, proteome, and a limited epigenome of ESCs following shRNA-enforced Nanog depletion [30]. Surprisingly, there was a stunning lack of concordance between RNA and protein levels at many of the genes analyzed following Nanog depletion. Approximately fifty percent of genes with elevated or depressed expression levels did not result in corresponding protein changes. Examination of Pol II binding and H3K9/14ac enrichment further signaled that post-transcriptional and epigenetic phenomena had significant and sometimes nonparallel effects on expression of different classes of genes. This study highlighted the importance of coupling proteomic analyses with chromatin occupancy and transcriptome profiling for a more comprehensive readout of gene expression changes in pluripotent cells.

5 RNA Polymerase II Transcriptional Machinery and the Mediator Coactivator

The regulation of gene expression is a highly complex process that requires heritable DNA-encoded nuclear factors, as well as epigenetic phenomena. In addition to the basal RNA Pol II machinery and associated general transcription factors (GTF) TFIIA, TFIIB, TFIID, TFIIE, TFIIF, and TFIIH, transcriptional coactivators assemble in complexes that modulate the frequency, specificity, and strength of gene expression. The Pol II TFIID contains TATA-binding protein (TBP) at its core and constitutes a complex of 13 or 14 TBP-associated factors (TAFs) that appear to differentially regulate different sets of target genes depending on subunit assembly and stoichiometry. TAF3 expression is essential for early vertebrate development and ESC-directed hematopoiesis [31], suggesting a developmental and tissue-specific requirement for some of the TAFs. Interestingly, TAF3 can function to anchor TFIID to nucleosomes through its PHD finger domain which is competent to bind trimethylated histone H3K4 [32]. Additionally, TAF1 has a double bromodomain that can bind diacetylated histone H4 [33]. Both modifications are marks of active chromatin. This was an unexpected discovery since prevailing views have long held that transcription takes place on promoters that are largely or completely devoid of histones. These recent data suggest a reshaping of this paradigm. Perhaps there is a threshold of low-level histone enrichment that allows transcriptionally permissive DNA that is loosely bound by nucleosomes to effectively dock with TFIID histone-binding components of the Pol II machinery. Alternatively, interaction of the Pol II machinery through permissive histone marks facilitates recruitment of chromatin remodeling machinery (discussed below) for the maximally effective dismantling of nucleosomes and the exposure of underlying DNA for transcription. In either scenario it is clear that these two regulatory layers may collaborate more closely than once thought to control gene expression.

Transcriptional activation by the Pol II machinery is facilitated by loading of the GTFs but high-level activation requires a 30-subunit coactivator known as the Mediator complex. A highly diverse protein module with four functionally separable domains, Mediator acts as a scaffold or a bridge coordinating interactions between Pol II and a wide range of coactivator complexes. Subunit heterogeneity within Mediator likely enables tissue and developmental specificity to the transcriptional machinery. Incorporation of different paralogs or substoichiometric levels of a particular subunit can confer precision to Mediator and the basal transcriptional machinery. Young and colleagues employed a small hairpin RNA interfence screen in V6.5 ESCs and revealed that downregulation of subunits within all four domains of Mediator contributed to a compromise in the ESC state [20]. In order of phenotypic severity, Med14, Med12, Med15, Med17, Med10, Med21, Med7, and Med6 were revealed in the screen. Of these, Med12 and Med21 are known to have an embryonic lethality phenotype if disrupted, corroborating the importance of these two subunits for pluripotency. It is unclear if other subunits will have significant developmental specification contributions since they have not been genetically disrupted in vivo or in ESC lines. Kagey et al. used ChIP-Seq to further demonstrate that Med1, Med12, TBP, Pol II, and the core pluripotency factors Oct4, Sox2, and Nanog were coenriched at many active promoters within ESCs. Intriguingly, Med12 coimmunoprecipitates with Nanog and regulates the pluripotency gene signature through Nanog [34]. Independent depletion of Med12 or Nanog resulted in a 50 % overlap between gene sets suggesting substantial functional overlap between the two proteins.

6 Histone Modifying and Chromatin Remodeling Complexes

The transcriptional response is modified by nucleosomal occupancy at gene promoters that characterizes whether DNA is accessible for transcription. Histones packaged within nucleosomes are subject to a wide array of posttranslational modifications on numerous residues [35, 36]. The modified histones likely serve collectively as scaffolds or docking sites for Pol II, the GTFs, transcriptional coactivators that include p300, Mediator, chromatin remodelers, and transcription factors. The pattern of histone modification, together with the frequency of methylated cytosine–guanine nucleotide pairs (CpG) at a promoter, contributes significantly to whether a chromatin domain will successfully attract the full repertoire of transcriptional machinery required for gene expression.

The employment of ChIP-chip and ChIP-Seq methods to analyze global histone enrichment has contributed significantly to a more comprehensive understanding of the transcriptional landscape in ESCs. The construction of genome-wide maps of histone modifications in ESCs, neural progenitor cells (NPC), and murine embryonic fibroblasts (MEF) has enabled cellular state and lineage specification potential to be ascertained largely from a histone signature. Bernstein and colleagues used ChIP-Seq to determine enrichment of trimethylated histone H3 residues located at positions lysine 4 (H3K4me3), H3K9me3, H3K27me3, H3K36me3, as well as histone H3 lysine 20 (H4K20me3) and RNA polymerase II (Pol2) [37]. The study was able to corroborate previous work by others demonstrating that H3K4me3 signals a permissive chromatin environment and the transcriptional start site at most genes, with the level of H3K4me3 often correlating with expression. Also shown on a global scale is that H3K27me3 marks repressive gene domains and H3K36me3 paints the bodies of coding and noncoding transcripts. These modifications were able to efficiently discriminate genes that are repressed, active, or in a poised state primed for activation in the three different cell types. This work built on sequencing of H3K4me3 and H3K27me3 at highly conserved regions of the genome by the same group and the discovery of "bivalent domains" enriched with both marks [38]. This coenrichment occurred most frequently at sites encoding transcription factors and other developmental regulators that are expressed at low levels in ESCs but become activated in a lineage-specific pattern during development. The current dogma is that genes that are simultaneously enriched with an activating and repressive mark enable rapid mobilization of gene expression at the appropriate time and place during development to control lineage specification. Not unexpectedly, ESCs have a greater percentage of bivalent domains than NPC or MEF, which is in concordance with ESCs having a far wider inventory of cellular lineages to specify during the developmental process.

Studies of chromatin state in human cells have mirrored observations made in their murine counterparts. ChIP-Seq analyses of 39 histone modifications in human T cells revealed a permissive histone signature that was highly predictive of active genes [39]. The 17 modifications that were enriched at over 3,000 promoters included several well-known activating marks that normally signal active gene expression such as histone H3 lysine 4 trimethylation (H3K4me3) and H3K9 acetylation were enriched at a significant proportion of promoters and gene bodies. Conversely, these active promoter regions and expressed genes were devoid of the repressive H3K27me3 mark. Additionally, a large collection of intergenic DNAseI hypersensitive sites obtained by high-throughput sequencing [40] were used to map and corroborate predicted enhancers, and a reliable histone-marking footprint for these *cis*-regulatory sites emerged. Though not always predictive of active gene expression, the experiments from Zhao and colleagues significantly extend our understanding of the highly combinatorial marking of histones and how this histone signature marks regulatory regions and may serve as a molecular beacon for transcriptional activation.

As discussed above, gene expression is largely dependent on co-assembly of the RNAPII transcriptional machinery and the Mediator complex at target promoters. The simplest models postulate that these core complexes collaborate with tissue and developmental specific as well as ubiquitous transcription factors to enable gene activation. However, a permissive chromatin environment is required for these complexes to engage their DNA targets. Nucleosomes must be removed and chromatin remodeled for naked DNA to be accessed. The mammalian SWI/SNF [also called BAF (Brg/Brahma-associated factors)] chromatin remodeling complexes are ATP-dependent machines that facilitate the remodeling of nucleosomes and serve as tissue-specific regulators of gene expression. A specialized BAF complex appears to be operative in maintaining self-renewal and pluripotency in ESCs and the early embryo. Proteomic studies have elucidated an ESC-specific BAF (termed esBAF) that has a defined subunit composition that differs from that found in neuronal cells or fibroblasts [41]. A recurring theme, this is reminiscent of the combinatorial composition observed in TFIID and Mediator complexes found in different tissues and developmental timepoints. An association of the core BAF component Brg with the core pluripotency factors Oct4, Sox2, and Sall4 was shown by Crabtree and colleagues. Our group also demonstrated that BAF155 and SMARCAD1—another SWI/SNF-related protein—existed within the pluripotency interactome network [12].

7 MicroRNAs and Gene Regulation

MicroRNAs are essential for ESC self-renewal and lineage specification [42]. These small noncoding RNAs exert their function by targeting the 3′ UTRs or coding sequences [43] of their RNA target sites resulting in the downregulation of gene expression. Although certain microRNAs are upregulated in ESCs [44, 45], how these microRNAs genetically intersect with the ESC transcriptional machinery has not been well characterized. Marson et al.

used H3K4me3 and H3K36me3 histone signatures to identify the putative promoter regions of 185 primary microRNA targets responsible for specifying over 300 mature microRNAs in murine and human ESCs, as well as in lineage-restricted cell types (MEFs and NPCs) [45].

This low-resolution analysis of global microRNA regulatory regions provides a global resource for determining how ESC-specific microRNAs are regulated. Confirming previous screening studies, Marson et al. used high-throughput RNA sequencing to analyze expression of all annotated microRNAs, including the ESC-enriched microRNA clusters mmu-mir-290–295 and mmu-mir-302–367. Following depletion of Oct4 using the doxycycline-inducible knockout ESC line ZHBTc4, expression of mature miRNAs from these clusters was downregulated modestly. Our own unpublished data using Taqman-based Q-PCR in Oct4-depleted ZHBTc4 ESCs indicates that levels of the mmu-mir-302–367 cluster may be higher in these cells and that downregulation of the mature miRNAs encoded by this cluster following Oct4 depletion is more significant. Additionally, both our study and that of Marson et al. show a downregulation of mir-135b following Oct4 depletion. Although expression of this microRNA is not confined to pluripotent tissues like the aforementioned clusters, it is occupied by more of the core and accessory pluripotency factors than most ESC-specific microRNAs (our unpublished observations). Additionally, it is highly expressed in human tumor populations [46]. Although contested in the literature [47, 48], many tumor populations appear to share differing semblances of an ESC pluripotency signature [47, 48]. Employing profiling strategies such as these will be useful following controlled differentiation of ESCs into hematopoietic progenitors. This will aid in the discovery of microRNAs that are essential for hematopoietic lineage specification.

Acknowledgment

Funding for this work was supported in part by a March of Dimes Basil O'Connor Scholar Award grant (5-FY10-45).

References

1. Evans MJ, Kaufman MH (1981) Establishment in culture of pluripotential cells from mouse embryos. Nature 292:154–156
2. Smithies O, Gregg RG, Boggs SS, Koralewski MA, Kucherlapati RS (1985) Insertion of DNA sequences into the human chromosomal beta-globin locus by homologous recombination. Nature 317:230–234
3. Doetschman T, Gregg RG, Maeda N, Hooper ML, Melton DW, Thompson S, Smithies O (1987) Targeted correction of a mutant HPRT gene in mouse embryonic stem cells. Nature 330:576–578
4. Hanna J, Markoulaki S, Schorderet P, Carey BW, Beard C, Wernig M, Creyghton MP, Steine EJ, Cassady JP, Foreman R et al (2008)

Direct reprogramming of terminally differentiated mature B lymphocytes to pluripotency. Cell 133:250–264
5. Okita K, Ichisaka T, Yamanaka S (2007) Generation of germline-competent induced pluripotent stem cells. Nature 448:313–317
6. Takahashi K, Yamanaka S (2006) Induction of pluripotent stem cells from mouse embryonic and adult fibroblast cultures by defined factors. Cell 126:663–676
7. Yu J, Vodyanik MA, Smuga-Otto K, Antosiewicz-Bourget J, Frane JL, Tian S, Nie J, Jonsdottir GA, Ruotti V, Stewart R et al (2007) Induced pluripotent stem cell lines derived from human somatic cells. Science 318:1917–1920
8. Ivanova N, Dobrin R, Lu R, Kotenko I, Levorse J, DeCoste C, Schafer X, Lun Y, Lemischka IR (2006) Dissecting self-renewal in stem cells with RNA interference. Nature 442:533–538
9. Matoba R, Niwa H, Masui S, Ohtsuka S, Carter MG, Sharov AA, Ko MS (2006) Dissecting Oct3/4-regulated gene networks in embryonic stem cells by expression profiling. PLoS One 1:e26
10. Chen X, Xu H, Yuan P, Fang F, Huss M, Vega VB, Wong E, Orlov YL, Zhang W, Jiang J et al (2008) Integration of external signaling pathways with the core transcriptional network in embryonic stem cells. Cell 133:1106–1117
11. Kim J, Chu J, Shen X, Wang J, Orkin SH (2008) An extended transcriptional network for pluripotency of embryonic stem cells. Cell 132:1049–1061
12. Wang J, Rao S, Chu J, Shen X, Levasseur DN, Theunissen TW, Orkin SH (2006) A protein interaction network for pluripotency of embryonic stem cells. Nature 444:364–368
13. Scholer HR, Ruppert S, Suzuki N, Chowdhury K, Gruss P (1990) New type of POU domain in germ line-specific protein Oct-4. Nature 344:435–439
14. Niwa H, Miyazaki J, Smith AG (2000) Quantitative expression of Oct-3/4 defines differentiation, dedifferentiation or self-renewal of ES cells. Nat Genet 24:372–376
15. Avilion AA, Nicolis SK, Pevny LH, Perez L, Vivian N, Lovell-Badge R (2003) Multipotent cell lineages in early mouse development depend on SOX2 function. Genes Dev 17:126–140
16. Kim JB, Sebastiano V, Wu G, Arauzo-Bravo MJ, Sasse P, Gentile L, Ko K, Ruau D, Ehrich M, van den Boom D et al (2009) Oct4-induced pluripotency in adult neural stem cells. Cell 136:411–419
17. Masui S, Nakatake Y, Toyooka Y, Shimosato D, Yagi R, Takahashi K, Okochi H, Okuda A, Matoba R, Sharov AA et al (2007) Pluripotency governed by Sox2 via regulation of Oct3/4 expression in mouse embryonic stem cells. Nat Cell Biol 9:625–635
18. Chambers I, Colby D, Robertson M, Nichols J, Lee S, Tweedie S, Smith A (2003) Functional expression cloning of Nanog, a pluripotency sustaining factor in embryonic stem cells. Cell 113:643–655
19. Mitsui K, Tokuzawa Y, Itoh H, Segawa K, Murakami M, Takahashi K, Maruyama M, Maeda M, Yamanaka S (2003) The homeoprotein Nanog is required for maintenance of pluripotency in mouse epiblast and ES cells. Cell 113:631–642
20. Kagey MH, Newman JJ, Bilodeau S, Zhan Y, Orlando DA, van Berkum NL, Ebmeier CC, Goossens J, Rahl PB, Levine SS et al (2010) Mediator and cohesin connect gene expression and chromatin architecture. Nature 467: 430–435
21. Heng JC, Feng B, Han J, Jiang J, Kraus P, Ng JH, Orlov YL, Huss M, Yang L, Lufkin T et al (2010) The nuclear receptor Nr5a2 can replace Oct4 in the reprogramming of murine somatic cells to pluripotent cells. Cell Stem Cell 6:167–174
22. Gu P, Goodwin B, Chung AC, Xu X, Wheeler DA, Price RR, Galardi C, Peng L, Latour AM, Koller BH et al (2005) Orphan nuclear receptor LRH-1 is required to maintain Oct4 expression at the epiblast stage of embryonic development. Mol Cell Biol 25:3492–3505
23. Rao S, Zhen S, Roumiantsev S, McDonald LT, Yuan GC, Orkin SH (2010) Differential roles of Sall4 isoforms in embryonic stem cell pluripotency. Mol Cell Biol 30:5364–5380
24. McKinney-Freeman SL, Lengerke C, Jang IH, Schmitt S, Wang Y, Philitas M, Shea J, Daley GQ (2008) Modulation of murine embryonic stem cell-derived CD41+c-kit+ hematopoietic progenitors by ectopic expression of Cdx genes. Blood 111:4944–4953
25. Boyer LA, Lee TI, Cole MF, Johnstone SE, Levine SS, Zucker JP, Guenther MG, Kumar RM, Murray HL, Jenner RG et al (2005) Core transcriptional regulatory circuitry in human embryonic stem cells. Cell 122:947–956
26. Hu G, Kim J, Xu Q, Leng Y, Orkin SH, Elledge SJ (2009) A genome-wide RNAi screen identifies a new transcriptional module required for self-renewal. Genes Dev 23:837–848
27. Loh YH, Wu Q, Chew JL, Vega VB, Zhang W, Chen X, Bourque G, George J, Leong B, Liu J et al (2006) The Oct4 and Nanog transcription network regulates pluripotency in mouse embryonic stem cells. Nat Genet 38: 431–440
28. Pardo M, Lang B, Yu L, Prosser H, Bradley A, Babu MM, Choudhary J (2010) An expanded Oct4 interaction network: implications for stem cell biology, development, and disease. Cell Stem Cell 6:382–395

29. van den Berg DL, Snoek T, Mullin NP, Yates A, Bezstarosti K, Demmers J, Chambers I, Poot RA (2010) An Oct4-centered protein interaction network in embryonic stem cells. Cell Stem Cell 6:369–381
30. Lu R, Markowetz F, Unwin RD, Leek JT, Airoldi EM, MacArthur BD, Lachmann A, Rozov R, Ma'ayan A, Boyer LA et al (2009) Systems-level dynamic analyses of fate change in murine embryonic stem cells. Nature 462:358–362
31. Hart DO, Santra MK, Raha T, Green MR (2009) Selective interaction between Trf3 and Taf3 required for early development and hematopoiesis. Dev Dyn 238:2540–2549
32. Vermeulen M, Mulder KW, Denissov S, Pijnappel WW, van Schaik FM, Varier RA, Baltissen MP, Stunnenberg HG, Mann M, Timmers HT (2007) Selective anchoring of TFIID to nucleosomes by trimethylation of histone H3 lysine 4. Cell 131:58–69
33. Jacobson RH, Ladurner AG, King DS, Tjian R (2000) Structure and function of a human TAFII250 double bromodomain module. Science 288:1422–1425
34. Tutter AV, Kowalski MP, Baltus GA, Iourgenko V, Labow M, Li E, Kadam S (2009) Role for Med12 in regulation of Nanog and Nanog target genes. J Biol Chem 284:3709–3718
35. Jenuwein T, Allis CD (2001) Translating the histone code. Science 293:1074–1080
36. Turner BM (2007) Defining an epigenetic code. Nat Cell Biol 9:2–6
37. Mikkelsen TS, Ku M, Jaffe DB, Issac B, Lieberman E, Giannoukos G, Alvarez P, Brockman W, Kim TK, Koche RP et al (2007) Genome-wide maps of chromatin state in pluripotent and lineage-committed cells. Nature 448:553–560
38. Bernstein BE, Mikkelsen TS, Xie X, Kamal M, Huebert DJ, Cuff J, Fry B, Meissner A, Wernig M, Plath K et al (2006) A bivalent chromatin structure marks key developmental genes in embryonic stem cells. Cell 125:315–326
39. Wang Z, Zang C, Rosenfeld JA, Schones DE, Barski A, Cuddapah S, Cui K, Roh TY, Peng W, Zhang MQ et al (2008) Combinatorial patterns of histone acetylations and methylations in the human genome. Nat Genet 40: 897–903
40. Crawford GE, Holt IE, Whittle J, Webb BD, Tai D, Davis S, Margulies EH, Chen Y, Bernat JA, Ginsburg D et al (2006) Genome-wide mapping of DNase hypersensitive sites using massively parallel signature sequencing (MPSS). Genome Res 16:123–131
41. Ho L, Ronan JL, Wu J, Staahl BT, Chen L, Kuo A, Lessard J, Nesvizhskii AI, Ranish J, Crabtree GR (2009) An embryonic stem cell chromatin remodeling complex, esBAF, is essential for embryonic stem cell self-renewal and pluripotency. Proc Natl Acad Sci U S A 106:5181–5186
42. Wang Y, Medvid R, Melton C, Jaenisch R, Blelloch R (2007) DGCR8 is essential for microRNA biogenesis and silencing of embryonic stem cell self-renewal. Nat Genet 39: 380–385
43. Tay Y, Zhang J, Thomson AM, Lim B, Rigoutsos I (2008) MicroRNAs to Nanog, Oct4 and Sox2 coding regions modulate embryonic stem cell differentiation. Nature 455:1124–1128
44. Houbaviy HB, Murray MF, Sharp PA (2003) Embryonic stem cell-specific MicroRNAs. Dev Cell 5:351–358
45. Marson A, Levine SS, Cole MF, Frampton GM, Brambrink T, Johnstone S, Guenther MG, Johnston WK, Wernig M, Newman J et al (2008) Connecting microRNA genes to the core transcriptional regulatory circuitry of embryonic stem cells. Cell 134:521–533
46. Sarver AL, French AJ, Borralho PM, Thayanithy V, Oberg AL, Silverstein KA, Morlan BW, Riska SM, Boardman LA, Cunningham JM et al (2009) Human colon cancer profiles show differential microRNA expression depending on mismatch repair status and are characteristic of undifferentiated proliferative states. BMC Cancer 9:401
47. Ben-Porath I, Thomson MW, Carey VJ, Ge R, Bell GW, Regev A, Weinberg RA (2008) An embryonic stem cell-like gene expression signature in poorly differentiated aggressive human tumors. Nat Genet 40:499–507
48. Kim J, Woo AJ, Chu J, Snow JW, Fujiwara Y, Kim CG, Cantor AB, Orkin SH (2010) A Myc network accounts for similarities between embryonic stem and cancer cell transcription programs. Cell 143:313–324

INDEX

Nicholas Zavazava (ed.), *Embryonic Stem Cell Immunobiology: Methods and Protocols*, Methods in Molecular Biology, vol. 1029, DOI 10.1007/978-1-62703-478-4,

MIX
Papier aus verantwortungsvollen Quellen
Paper from responsible sources
FSC® C105338

If you have any concerns about our products,
you can contact us on
ProductSafety@springernature.com

In case Publisher is established outside the EU,
the EU authorized representative is:
Springer Nature Customer Service Center GmbH
Europaplatz 3, 69115 Heidelberg, Germany

Printed by Libri Plureos GmbH
in Hamburg, Germany